International Association of Fire Chiefs

EXAM PREP

Fire Inspector I & II

By Dr. Ben A. Hirst,
Performance Training
Systems

JONES AND BARTLETT PUBLISHERS
Sudbury, Massachusetts
BOSTON TORONTO LONDON SINGAPORE

Jones and Bartlett Publishers
World Headquarters
40 Tall Pine Drive
Sudbury, MA 01776
978-443-5000
www.jbpub.com

International Association of Fire Chiefs
4025 Fair Ridge Drive
Fairfax, VA 22033
www.IAFC.org

Performance Training Systems, Inc.
760 U.S. Highway One, Suite 101
North Palm Beach, FL 33408
www.FireTestBanks.com

Jones and Bartlett Publishers Canada
2406 Nikanna Road
Mississauga, ON L5C 2W6
Canada

Jones and Bartlett Publishers International
Barb House, Barb Mews
London W6 7PA
United Kingdom

Jones and Bartlett's books and products are available through most bookstores and online booksellers. To contact Jones and Bartlett Publishers directly, call 800-832-0034, fax 978-443-8000, or visit our website www.jbpub.com.

Substantial discounts on bulk quantities of Jones and Bartlett's publications are available to corporations, professional associations, and other qualified organizations. For details and specific discount information, contact the special sales department at Jones and Bartlett via the above contact information or send an email to specialsales@jbpub.com.

Editorial Credits

Author: Dr. Ben A. Hirst

Production Credits

Chief Executive Officer: Clayton E. Jones
Chief Operating Officer: Donald W. Jones, Jr.
President: Robert W. Holland, Jr.
V.P., Sales and Marketing: William J. Kane
V.P., Production and Design: Anne Spencer
V.P., Manufacturing and Inventory Control: Therese Bräuer
Publisher, Public Safety Group: Kimberly BrophyAssociate
Production Editor: Karen C. Ferreira
Editorial Assistant: Adrienne Zicht
Director of Marketing: Alisha Weisman
Cover Design: Kristin E. Ohlin
Interior Design: Anne Spencer
Illustration: George Nichols
Composition: Northeast Compositors
Printing and Binding: Courier Stoughton

Photo Credits

Photographs of wet barrel hydrant (pages 27 and 46) and dry barrel hydrant (page 29) courtesy of American AVK Company.
Cover photograph courtesy of Dennis Wetherhold, Jr.

ISBN: 0-7637-2848-9

The procedures in this text are based on the most current recommendations of responsible sources. The publisher and Performance Training Sysyems, Inc. make no guarantees as to, and assume no responsibility for the correctness, sufficiency, or completeness of such information or recommendations. Other or additional safety measures may be required under particular circumstances. This text is intended solely as a guide to the appropriate procedures to be employed when responding to an emergency. It is not intended as a statement of the procedures required in any particular situation, because circumstances can vary widely from one emergency to another. Nor is it intended that this text shall in any way advise firefighting personnel concerning legal authority to perform the activities or procedures discussed. Such local determination should be made only with the aid of legal counsel.

Printed in the United States of America
09 08 07 06 05 04 10 9 8 7 6 5 4 3 2 1

CONTENTS

ACKNOWLEDGEMENTS

More than 90 fire department Fire Inspectors have contributed to the development, validation, revision, and updating of the test items included in this Exam Prep book. Their efforts have spanned more than fifteen years and are valued because of the credibility they provided. A special thanks goes to the recent Technical Review Committees for Fire Inspector I and II for validating and updating the test items to the latest NFPA Standard and latest technical publications: David Collado, Palm Beach County Fire and Rescue, Florida; Brian Sauer, Pike Township Fire Department, Indiana; Keith Frazier, Arkansas Fire Academy, Arkansas; John Powers, Minnesota Fire Service Certification Board, Minnesota; Ron Lovett, Indiana Public Service Training Institute, Indiana; Randal Frisinger, Texas Commission on Fire Protection, Texas; and Scott Spencer, Utah Fire and Rescue, Utah. These individuals worked diligently to make considerable improvements in the test items.

I want to thank my wife Elizabeth, family, and friends who encouraged me to continue pressing forward with the work. Without their understanding and support I would not have been able to meet the scheduled delivery.

Last, but not least, I express my sincere thanks to my able staff: Ellen Korn, Administrative Assistant; Diane Mattis, Sales Support; Leigh Kaufman, Editorial Support; Walter Hirst, Paramedic/Firefighter and Director of Operations; Todd Lynch, Paramedic/Firefighter and Regional Sales Manager; Chris Seay, Paramedic/Firefighter and Regional Sales Manager; Ed Beardsley, Paramedic/Firefighter and Regional Sales Manager; Joseph Mazzeo, Paramedic/Firefighter and Regional Sales Representative. While I was away, in complete solitude, they kept the business going. Ellen Korn is due special recognition because she made sure everything in this book was accurate and formatted properly, and that the office ran smoothly in my absence.

PREFACE

The Fire and Emergency Medical Service is facing one of the most challenging periods in its history. Local, state, provincial, national, and international government organizations are under pressure to deliver ever-increasing services. The events of September 11, 2001, continued activities and threats by terrorist organizations worldwide, and the need to maximize available funds are some of the reasons causing Fire and Emergency Medical Service organizations to examine and reinvent their roles.

The challenge of reinventing the Fire and Emergency Medical Service to provide the first response efforts includes increasing professional requirements. Organizations such as the National Fire Protection Association (NFPA), National Professional Qualifications Board (Pro Board), International Fire Service Accreditation Congress (IFSAC), International Association of Fire Chiefs (IAFC), and International Association of Fire Fighters (IAFF) are dramatically affecting the field by raising the professional qualifications of the first line of defense for emergency response.

Qualification standards have clearly been improved. Accreditation of training and certification are at the highest levels ever in the history of the Fire and Emergency Medical Service. These improvements are reflected in a better prepared first responder, albeit not without affecting those individuals who serve. Firefighters are being required to expand their roles, acquire new knowledge, develop new and higher-level technical skills, and participate in requalification and in-service training programs on a regular basis.

The aftermath of September 11, 2001, has had a profound effect on the Fire and Emergency Medical Service. Lessons learned, new technology, and a national focus on terrorism and weapons of mass destruction are placing much greater demands on firefighters to keep abreast of changes in their specialty operations and to improve their technical competence in new technology that was not available even just a few years ago.

Firefighters cannot afford to be complacent and continue to perform in the same way. The obvious dangers faced by first responders under today's heightened security conditions require many adjustments in what is being taught to firefighters as they operate in an emergency environment. Processes and modes of operation must be carefully examined and must be continuously monitored, changed, and updated.

National leaders constantly point out the crucial role played by first responders as the "first line of defense" against acts of terror and defense of life and property from extremely dangerous weapons that have never been used extensively in U.S. history. Firefighting is highly steeped in tradition. One critical issue is the need to question our traditions and our traditional thinking to bring our knowledge, skills, and abilities in line with the demands of today's real world.

Many things have been learned from the September 11 terrorist attacks on the United States. Some of these lessons resulted from our reluctance to change processes and procedures (i.e., our traditions). As great as those traditions are, members of the Fire and Emergency Medical Service must not stop reflecting on the paramount reasons for our existence: to protect property, to save lives, and to perform our tasks with personal safety as the number one concern. These are very important reasons to exist, to improve, and to move from a good fire and emergency service industry to a great fire and emergency service industry.

A word about the importance of strategic and tactical information. Many organizations focus much of their training time and effort on the performance side of firefighting. That endeavor is essential and is the bottom line for developing skilled firefighters. The dark side of this approach to training is its failure to emphasize key knowledge requirements. Often, it is not what we did or didn't do as firefighters, but rather what we could have done with a strong base of knowledge that helps to analyze and detect a need for action outside the routine tasks of firefighting. How many people would have been saved on September 11 if a unified command had ordered an immediate evacuation of the World Trade Center towers and surrounding structures? How many people died because they were sent back into the structures by first responders on the scene? How many died after being directed to go to the roofs of the towers and other buildings for rescue? Why weren't the basic principles of apparatus placement followed? Could the entire situation have been handled better if more emphasis had been placed on knowledge of building construction, collapse zones, nearby safe havens, and better placement and location of key apparatus and communications equipment? These questions may never be answered, but they do require the Fire and Emergency Medical Service as a whole, and each firefighter as an individual, to focus equally on the knowledge portion of firefighting to help improve the performance side of our tasks.

We are not training robots or race horses. Our fire officers, firefighters, and support personnel must develop a solid knowledge base so that better judgments, sizeups, and fire ground actions become possible. Research in education and training over the years has concluded that lack of knowledge is one of the key reasons why tasks are poorly performed or performed in a manner that did not achieve the expected results.

Today, hazardous materials are found everywhere in U.S. society. Homes contain them, as do businesses, warehouses, places of public assembly, and almost every other type of structure. These materials can be found in open areas such as farm land, wooded areas where clandestine labs operate, and countless other outdoor areas. First responders must learn to treat fires, collapses, and even traffic accidents as probable hazardous materials incidents.

There is a great deal to know and master regarding the nine hazard classes identified and described by the U.S. Department of Transportation (DOT). Many hazardous materials incidents occur in the United States every day. First responders must realize that their personal safety and the safety of citizens are paramount during the response to a hazardous materials incident. Firefighter safety is, of course, always the first priority during fire ground operations. This realization is extremely critical when hazardous materials are detected or the preplan suggests that they are present at the incident. Proactive planning can aid first responders in efficiently and effectively handling critical hazardous materials incidents. One of the most harmful situations is the discovery of a hazardous material well into the response situation. The unexpected often causes undue exposure to harmful substances that can have both immediate and long-term health effects. The first responder and the initial hazardous materials operations crew must have highly developed recognition skills and identification skills, and must know the isolation and protection parameters required for the incident.

Never in the history of the Fire and Emergency Medical Service has it become so clear that learning is a careerlong, lifelong requirement for coping with the demands that lie ahead. Members of the Fire and Emergency Medical Service must adopt this principle to move from our tradition-rich past to become truly great providers and protectors available to everyone we serve. We cannot correct the mistakes of the past, but we can use lessons learned to prevent similar mistakes in the future. Knowledge is power. Efficient and effective people are the solution to moving from a good Fire and Emergency Medical Service to a truly great one.

Fire inspection, code enforcement, and public education have long been recognized as the primary ways to help prevent fires, protect property, and save lives. Even though the Fire and Emergency Medical Service has been well aware of this lower-cost approach to prevent fires, much remains to be accomplished. The United States leads the industrialized nations of the world in fire-related damage and deaths, according to the Recommissioned Panel for *America Burning* (May 2000). Even more appalling is the fact that the loss of life and property is in the billions of dollars, totaling more than the annual losses from floods, hurricanes, tornadoes, earthquakes, and all other natural disasters combined (*Fire in the United States*, eleventh edition, August 1999).

Why is this happening? It appears that Americans just don't know about these conditions. Fire Inspectors, Plan Examiners, and Fire and Life Safety Educators must do a better job in prevention activities and public education. Many Fire and Emergency Medical Service organizations will blame the lack of funds as the main barrier to reducing these tremendous losses. That could be true—but only up to a point.

It is time to bite the bullet! Members of the Fire and Emergency Medical Service community must do a better job in getting the word out about the savings that can result from improved fire prevention programs. These programs are far less expensive to the community when viewed in comparison to fire suppression costs.

More fire departments need to be proactive in improving their fire inspection, code enforcement, public education, and public information programs to address specific community needs. Far too few fire departments have well-developed public education programs. Most of the really good programs are found in the communities that can afford them instead of the communities that have the highest instances of fire loss. Achieving this goal may require national budget considerations, a focus of time and personnel on other community-related projects dealing with fire prevention and public education, or some combination of these approaches.

Does your fire department have a well-organized public fire prevention and education program? Are members of the Fire and Emergency Medical Service devoting enough time to identifying and correcting fire hazards in our community? Do we place fire prevention and education programs as high on our priority list as we do fire suppression and response times? Probably not! If we were more effective, property losses and human lives lost would be declining each year rather than increasing. According to data in *America Burning*, literally billions of dollars could be saved every two to three years by concentrating more effort and funding on fire prevention methods and public education than by pursuing any other approach.

This *Fire Inspector Exam Prep* was developed to help ensure that Fire Inspectors will be more knowledgeable, more aware of their professional responsibilities, better informed about their legal liabilities, and better equipped with information that can protect their community property and save lives.

A fire can do much more than damage a structure. It can cause economic chaos in a community. Consider a fire in an industrial complex that employs many residents in the community. Such a fire can cause a loss of jobs in the industry and local business establishments and undercut economic gains made over years of community effort. If the fire resulted from some known code violation, then the impact can have huge liability implications for the local fire jurisdiction.

Fire and Emergency Medical Service members generally don't like to take examinations. (Actually, few people really like them.) The primary purposes of the *Exam Prep* series are to help Fire and Emergency Medical Service personnel improve their level of knowledge, eliminate examination-taking fear, build self confidence, and develop good study and information mastery skills.

Performance Training Systems, Inc. (PTS), has emerged over the past 16 years as the leading provider of valid examination materials for certification, promotion, and training for fire and emergency medical personnel. More than 30 examination-item banks provide the basis for validated examinations. All products are based on the NFPA Professional Qualifications Standards and the DOT Curriculum for Medical First Responders.

Over the past seven years, PTS has conducted research supporting the development of the Systematic Approach to Examination Preparation® (SAEP). SAEP® has resulted in consistent improvement in scores for persons taking certification, promotion, and training completion examinations. This *Exam Prep* book is designed to assist firefighters in improving their knowledge, skills, and abilities while seeking training program completion, certification, and promotion. SAEP, coupled with helpful examination-taking tips and hints, will help ensure improved performance from a more knowledgeable and skilled firefighter.

All examination questions used in SAEP were written by fire and emergency service personnel; technical content was validated through the use of current technical textbooks and other technical reference materials; and job content was validated by technical review committees representative of the specific jobs in the fire service, training, and certification organizations. The examination questions in the *Exam Prep* series represent an approximate 60 percent sample of the various examination-item banks developed and maintained by PTS over the past 16 years. These examination materials are being used by 60 fire service certification agencies worldwide, 106 fire academies, and more than 300 fire department training divisions. Forty-six of the 50 state fire service certification agencies use these examination materials in their programs of certification. For more information on the number of available examination banks and the processes of validation, visit *firetestbanks.com*.

Introduction to the Systematic Approach to Examination Preparation

How does SAEP work? SAEP is an organized process of carefully researched phases that permits each person to proceed in examination preparation at that individual's preferred pace. At certain points, self-study is required to move from one phase of the program to another. Feedback on progress is the basis of SAEP. It is important to follow the program steps carefully to realize the full benefits of the system.

SAEP allows you to prepare for your next comprehensive training, promotional, or certification examination. Just follow the steps to success. Performance Training Systems, Inc. (PTS), the leader in producing promotional and certification examinations for the Fire and Emergency Medical Service industry, and the developer of SAEP, has the experience and examination expertise needed to help you meet your professional goals.

Taking the preparatory examinations will enable you to pinpoint your areas of weakness in terms of NFPA Standard 1031, and the feedback provided will include the reference and page number to help you research the questions that you miss or guess using current technical reference materials. This program is a three-examination set for Fire Inspector I and Fire Inspector II as described in *NFPA 1031, Standard for Professional Qualifications for Fire Inspector and Plan Examiner,* 2003 edition.

Primary benefits of the SAEP include the following:

- Emphasis on areas of weakness
- Immediate feedback
- Savings in terms of time and energy
- Learning technical material through context and association
- Helpful examination preparation practices and hints

SAEP is organized in four distinct phases for the Fire Inspector I and II certifications described in NFPA 1031. The phases are briefly described next.

Phase I

Phase I includes three examinations containing items that are selected from each major part of *NFPA 1031, Standard for Professional Qualifications for Fire Inspector and Plan Examiner,* 2003 edition.

An essential part of the SAEP design is to survey your present level of knowledge and then build on that base for subsequent examination and self-directed study activities. Therefore, it is suggested that you read the reference materials but do not study or look up any answers while taking the initial examination. Upon completion of the initial examination, you will complete a feedback activity and record examination items that you missed or that you guessed. Once you have completed the initial examination and researched the answers for any questions you missed, you may proceed to the next examination. This process is repeated throughout the Fire Inspector I and II series, depending on the level of certification you are seeking.

Phase II

Fire Inspector II examinations are provided in SAEP Phase II. This phase includes three examinations, each made up of examination items from appropriate sections of *NFPA 1031, Standard for Professional Qualifications for Fire Inspector and Plan Examiner,* 2003 edition.

The examinations should be completed as prescribed in the directions supplied with the examination. Complete the feedback report using the procedures outlined in the answer and feedback section. Pay particular attention to those areas of the references covering material where you score the lowest. At this point, you should read the materials containing the correct response in context once again. This technique will help you master the material, relate it to other important information, and retain knowledge.

Phase III

Phase III contains important information about examination-item construction. It provides insight regarding the examination-item developers, the way they apply their technology, and hints and tips to help you score higher on any examination. Read this phase carefully. It is a good practice to read it twice and to study the information it contains a day or two prior to your scheduled examination.

Phase IV

Phase IV information addresses the mental and physical aspects of examination preparation. Do not skip this part of your preparation. Points can be lost if you are not ready, physically and mentally, for the examination. If you have participated in sports or other competitive events, you know the importance of this level of preparation. There is no substitute for readiness. Just being able to answer the questions will not move you to a level of excellence and bring you to the top of the examination list for training, promotion, or certification. Quality preparation involves much more than just answering examination items.

Supplemental Practice Examination Program

The supplemental practice examination program differs from the SAEP program in several ways. It is provided over the Internet 24 hours a day, 7 days a week. The supplemental practice examination allows you to make final preparations immediately before your examination date. You will get an immediate feedback report that includes the questions missed and the references and page numbers for those missed questions. The practice examination allows you to concentrate on your areas of greatest weakness and will save you time and energy immediately before the examination date. If you choose this method of preparation, do not "cram" for the examination. The helpful hints for examination preparation will explain the reasons for avoiding a "cramming exercise." A supplemental practice examination is available when you purchase this *Exam Prep* book and can be accessed by using the enclosed registration form. Do not forget to fax a copy of your Personal Progress Plotter along with your registration form. The data supplied on your Personal Progress Plotter will be kept confidential and will be used by PTS to make future improvements in the *Exam Prep* series. You may take a short practice examination to get the procedure clear in your mind at *www.webtesting.cc*.

Good luck in your efforts to improve your knowledge and skills. Our primary goal is to improve the Fire and Emergency Medical Service one person at a time. We want your feedback and impression of the system to help us implement improvements in future editions of the *Exam Prep* series. Address your comments and suggestions to *www.firetestbanks.com*.

Rule 1

Examination preparation is not easy. Preparation is 95 percent perspiration and 5 percent inspiration.

Rule 2

Follow the steps very carefully. Do not try to reinvent or shortcut the system. It really works just as it was designed to!

Personal Progress Plotter

Fire Inspector I Exam Prep

Name: ____________________

Date Started: ____________________

Date Completed: ____________________

Fire Inspector II Exam Prep

Name: ____________________

Date Started: ____________________

Date Completed: ____________________

Fire Inspector I	Number Guessed	Number Missed	Examination Score
Examination I-1			
Examination I-2			
Examination I-3			

Fire Inspector II	Number Guessed	Number Missed	Examination Score
Examination II-1			
Examination II-2			
Examination II-3			

Formula to compute Examination Score = 100 − ((Number guessed + Number missed) × Point Value per Examination Item).

Note: 150 Examination-item examination = 0.67 points per examination item

100 Examination-item examination = 1.0 points per examination item

75 Examination-item examination = 1.34 points per examination item

Example: Examination I-1, 5 examination items were guessed, 8 were missed, for a total of 13 on a 150-item examination. Examination score would be 100 − (13 × 0.67 point) = 91.3.

Example: Examination I-1, 5 examination items were guessed, 8 were missed, for a total of 13 on a 100 item examination. Examination score would be 100 − (13 × 1.0 point) = 87.0.

Example: Examination II-1, 5 examination items were guessed, 8 were missed, for a total of 13 on a 75-item examination. Examination score would be 100 − (13 × 1.34 points) = 82.6.

Note: To receive your free online practice examination, you must fax a copy of your completed Personal Progress Plotter along with your registration form.

PHASE I

Fire Inspector I

Examination I-1: Beginning NFPA Standard 1031

Taking the 100-Item Examination: Examination I-1. Read the reference materials but do not study prior to taking the examination. The examination is designed to identify your weakest areas in terms of NFPA Standard 1031. There will be steps in SAEP that require self-study of specific reference materials. Remove Examination I-1 from the book. Mark all answers in ink, which ensures that no changes are made. Do not mark through answers or change answers in any way once you have selected the answer.

Step 1—Take Examination I-1. When you have completed Examination I-1, compare your answers with the correct answers in Appendix A. Each answer cites relevant reference materials with page numbers. If you answered the examination item incorrectly, you have a source for conducting your correct answer research.

Step 2—Score Examination I-1. How many examination items did you miss? Write the number of missed examination items in the blank in ink ________________. Enter the number of examination items you guessed in this blank ________________. Enter these numbers in the designated locations in your Personal Progress Plotter.

Step 3—The learning begins! Carefully research the page cited in the reference material for the correct answer. For instance, use IFSTA, *Fire Inspection and Code Enforcement*, sixth edition, go to the page number provided, and find the answer.

Rule 3

Mark with an "X" any examination items for which you guessed the answer. To obtain the maximum return on effort, research any answer that you guessed even if you guessed correctly. Find the correct answer, highlight it, and then read the entire paragraph that contains the answer. Be honest and mark all questions on which you guessed. Some examinations have a correction for guessing built into the scoring process. The correction for guessing can reduce your final examination score. If you are guessing, you are not mastering the material.

Helpful Hint

Your first impression is often the best. More than 41 percent of answers changed during our SAEP field test were changed from a right answer to a wrong answer. Another 33 percent were changed from a wrong answer to another wrong answer. Only 26 percent of answers were changed from a wrong answer to a right answer. In fact, a number of changed answers resulted in the participant not making a perfect score of 100 percent! Think twice before you change your answer. The odds are not in your favor.

Helpful Hint

Researching correct answers is one of the most important activities in SAEP. Locate the correct answer for all missed examination items. Highlight the correct answer. Then read the entire paragraph containing the answer. This will put the answer in context for you and provide important learning by association.

Helpful Hint

Work through all missed examination items using the same technique. Reading the entire paragraph improves retention of the information and helps you develop an association with the material and learn the correct answers. This step may sound simple. A major finding during the development and field testing of SAEP was that you learn from your mistakes.

Examination I-1

Directions

Remove Examination I-1 from the manual. First, take a careful look at the examination. There should be 100 examination items. Notice that a blank line precedes each examination item number. Enter the answer to the examination item on this line. Write the answer in ink. Remember the rule about not changing your answers. Changed answers are often incorrect, and more often than not the answer that is chosen first is correct.

If you guess the answer to a question, place an "X" or a check mark by your answer. This step is vitally important as you gain and master knowledge. We will explain how we treat the "guessed" items later in SAEP.

Take the examination. Once you complete it, go to Appendix A and score your examination. Carefully follow the directions for feedback on the missed and guessed examination items.

_____ **1.** The first step in the enforcement procedure would **usually** be to:
A. notify the responsible party of his/her appeal rights.
B. determine whether the violation is significant enough in nature to pursue.
C. send the business owner a letter stating the premises are in full compliance.
D. notify the responsible party, in writing, of all violations found.

_____ **2.** Procedures in various jurisdictions may differ due to differences in the code enforcement each has adopted. These procedures should be:
A. restricted to a single typed page.
B. designed to promote appeals.
C. in detailed, written form.
D. developed in a way that can be easily changed.
E. maintained under strict security.

_____ **3.** Completed inspection reports are needed for every inspection because they:
A. may be used for future code enforcement.
B. are needed to complete the plan review.
C. are used during community zoning hearings.
D. indicate the effectiveness of the current inspection program.

_____ **4.** Which of the following **is not** helpful to consider when writing a letter?
A. The reader will be reading the letter with interest.
B. Use a natural and clear expression of thoughts.
C. Maintain an awareness of the thoughts, feelings, and impressions that might be evoked when the recipient reads the letter.
D. Remember not to think in terms of the letter being a substitute for an actual conversation.

_____ **5.** _________________ are a way of staying aware of changes in use or hazardous conditions.
A. Surveys
B. Permits
C. Self-inspections
D. Fire reports

_____ **6.** On a site plan, the north directional symbol **usually** points toward the:
A. magnetic North Pole.
B. right side of the page.
C. top of the page.
D. true North Pole.

_____ **7.** There are four main views of working drawings. They include all of the following **except** the _________________ view.
A. sectional
B. utility
C. elevation
D. detailed
E. plan

_____ **8.** When a citizen calls to report a suspected fire hazard, fire inspectors should:
A. record all pertinent information.
B. file the complaint for future action.
C. forward the complaint to the Board of Appeals if immediate corrective action is required.
D. consider the motivation for the complaint.

_____ **9.** When investigating a complaint, a fire inspector should:
A. never give the owner advance notice of the investigation.
B. explain the purpose of the investigation to the owner.
C. take appropriate departmental action to correct the violations.
D. Both B and C are correct.

_____ **10.** When an individual building or business owner feels that the local jurisdiction has made an improper code interpretation or decision, the individual may request a:
A. hearing in district court or civil court, depending on the degree of hardship.
B. hearing by the civilian review board.
C. review by the Board of Appeals.
D. more favorable ruling by the building inspector.

_____ **11.** Prior to a court case, an inspector should do all of the following **except**:
A. reinspect the facility the day before the trial.
B. go over his/her testimony with the prosecutor before entering the courtroom.
C. appear in proper uniform or be neatly dressed.
D. volunteer information to make responses as detailed and complete as possible.

_____ **12.** Which of the following **is not** suggested for courtroom procedure or behavior?
A. Never become argumentative on the witness stand.
B. Make sure that all physical evidence, exhibits, photographs, notes, and reference materials are brought to court.
C. Attempt to answer a question you do not know.
D. Remain impartial. Do not give the impression that you have a personal dislike for the defendant.

_____ **13.** According to NFPA 101, Life Safety Code, a building that provides sleeping facilities for four or more residents and is occupied by persons who are generally prevented from protecting themselves because of security measures not under their control, is classified as a/an:

A. residential occupancy.
B. occupancy of unusual structure.
C. industrial occupancy.
D. detention and correctional occupancy.

_____ **14.** According to NFPA 101, Life Safety Code, buildings used as a store, market, and other rooms, buildings, or structures used to display and sell merchandise are classified as a/an:

A. mercantile occupancy.
B. business occupancy.
C. storage occupancy.
D. place of assembly.
E. occupancy of unusual structures.

_____ **15.** Fire drills in educational occupancies **should not be** conducted:

A. during classroom changes, since the students are already in the hallways.
B. during assembly programs, since the students are all in one place making for easy evacuation.
C. during lunch time.
D. at the start of each day.

_____ **16.** The purpose of school fire exit drills is to:

A. increase speed in evacuating all children.
B. instill discipline and order.
C. ensure that all exits are being used and are not locked.
D. ensure orderly exit of all personnel under controlled supervision.

_____ **17.** A sketch depicting the general arrangement of the property in reference to streets, adjacent properties, and other important features is known as a:

A. plot plan.
B. sectional view.
C. floor plan.
D. blueprint.

_____ **18.** Which item **would not** be considered during a field observation of a site for emergency access?

A. Fire department access roadways which are wide enough to permit fire apparatus to operate and to pass
B. Verification of code compliance for the storage, handling, and use of hazardous materials
C. Physical, topographical, or architectural obstructions
D. Adequate turning radius
E. Whether access roads are within minimum building access distance requirements

_____ **19.** The temperature at which a liquid fuel, once ignited, will continue to burn is known as:
A. fire point.
B. vapor temperature.
C. boiling point.
D. flash point.

_____ **20.** When a fuel gives off enough vapors so that it can be ignited and burn momentarily, it has reached its:
A. fire point.
B. upper flammable limit.
C. vapor density.
D. flash point.

_____ **21.** Class B fires involve fuels such as:
A. greases or flammable/combustible liquids.
B. energized-electrical equipment.
C. combustible metals.
D. ordinary combustibles.

_____ **22.** When referring to the L.E.L. of a flammable or combustible liquid, what does the L.E.L. mean?
A. Lowest Efficiency Level
B. Life-Span Equivalency Label
C. Lower Explosive Limit
D. Lowest Evaporation Limit

_____ **23.** The upper and lower concentrations of a vapor that will produce a flame at a given pressure and temperature are called:
A. flash points.
B. vapor densities.
C. flammable and explosive limits.
D. burning points.

_____ **24.** The percentage of a flammable substance in air that will burn when in contact with an ignition source <u>**best**</u> defines:
A. ignition temperature.
B. boiling point.
C. flammable/explosive limit.
D. flash point.

_____ **25.** Vapor density is used to evaluate the relative weights of _______________ in much the same way as specific gravity is used to evaluate _______________.
A. solids, liquids
B. air, solids
C. gases, liquids
D. liquids, air

_____ **26.** Choose the <u>best</u> answer regarding the difference between compressed and liquefied gases at normal temperatures. Within a pressurized container:
A. compressed gases are in both a liquid and gaseous state.
B. compressed gases are only in a gaseous state.
C. liquefied gases are only in a liquid state.
D. liquefied gases are in a gaseous state only.

_____ **27.** In a closed area, gases with vapor density of less than one will:
A. rise and not concentrate near the ceiling.
B. rise and concentrate near the ceiling.
C. settle and not concentrate at the floor.
D. settle and concentrate at the floor.

_____ **28.** Vapor density is defined as the weight of ________________ as compared to the weight of an equal volume of ________________:
A. gas, air.
B. air, gas.
C. liquid, air.
D. air, liquid.

_____ **29.** On above-ground inside storage tanks for flammable or combustible liquid storage tanks, automatic closing, heat-actuated valves are required on:
A. all piping connections.
B. piping connections above the liquid level.
C. none of the piping connections.
D. all piping connections below the liquid level.

_____ **30.** Inside storage tanks containing Class I, II, or III-A liquids must have vent pipes which terminate:
A. outside of building.
B. at least 12 feet above tank.
C. at least 5 feet above tank.
D. at least 3 feet above fill pipe.

_____ **31.** Containers that are approved for flammable and combustible liquids should have <u>how many</u> devices that will provide sufficient venting capacity to limit the internal pressure of the containers to 10 psi or 30 percent of the bursting pressure of the container?
A. One or more
B. Three or more
C. Four or more
D. Not required

_____ **32.** The diked area surrounding above-ground flammable liquid storage tanks must have a volume large enough to contain:
A. the volume of half of the tanks within the diked area.
B. half of the volume of the largest tank.
C. entire volume that could be released from the largest tank.
D. the volume of all the tanks within the diked area.

_____ **33.** Dip tank operations must be located in noncombustible buildings. At what floor level in noncombustible buildings must dip tank operations be located?
A. Not above the third floor.
B. Not above the second floor.
C. Not below the ground floor.
D. Not above or below the ground floor.

_____ **34.** To avoid trapping cryogenic liquids in piping, which choice for installed piping is the safest?
A. A pressure relief device every 100 ft and piping sloped up from the container.
B. A pressure relief device between every two shut-off valves and piping sloped up from the container.
C. A pressure relief device every 90 ft and piping sloped down from the container.
D. A pressure relief device between every two shut-off valves and piping sloped down from the container.

_____ **35.** The objective of a residential inspection program is to:
A. attain proper life safety conditions.
B. prevent fires.
C. educate the property owner and occupants about fire safety.
D. All of the above.

_____ **36.** The colder of two bodies will always absorb heat until both objects are the same temperature is explained by the Law of ________________ which is a natural law of physics.
A. Specific Heat
B. Heat Transfer
C. Latent Heat of Vaporization
D. Atomic Weight

_____ **37.** The principle which <u>most</u> <u>closely</u> describes how water extinguishes fire is:
A. removal of fuel.
B. reduction of temperature.
C. exclusion of oxygen.
D. inhibition of a chain reaction.

_____ **38.** During the first phase of a fire, the oxygen concentration within the room contains approximately ________________ percent oxygen.
A. 10
B. 16
C. 20
D. 79

_____ **39.** A fire in the presence of a higher than normal concentration of oxygen will:
A. burn slower than normal.
B. burn faster than normal.
C. not be effected by the oxygen.
D. not burn if oxygen is too rich.

_____ **40.** _______________ is electrical heat energy resulting from inadequately insulated electrical materials.
A. Dielectric heating
B. Induction heating
C. Frictional heat
D. Leakage current heating

_____ **41.** Mechanical heat energy initiated by the movement between objects in contact with one another defines:
A. dielectric heating.
B. heat of friction.
C. static electricity.
D. resistance heating.

_____ **42.** The amount of heat generated through the process of oxidation is known as:
A. heat of solution.
B. heat of decomposition.
C. heat of combustion.
D. spontaneous heating.

_____ **43.** Heat that is released by the mixture of matter within certain liquids is known as:
A. heat of solution.
B. heat of decomposition.
C. heat of combustion.
D. spontaneous heating.

_____ **44.** Heat that results from passing an electrical force through a conductor defines _______________ heating.
A. resistance
B. dielectric
C. induction
D. frictional

_____ **45.** Heating of an organic substance without an external heat source is known as:
A. spontaneous heating.
B. heat of decomposition.
C. dielectric heating.
D. frictional heat.

_____ **46.** The release of heat usually due to bacterial action best defines:
A. spontaneous heating.
B. heat of decomposition.
C. heat of compression.
D. heat of solution.

_____ **47.** The three methods of heat transfer are:
A. conflagration, combustion, and contact.
B. fuel, heat, and oxygen.
C. conduction, radioactivity, and direct flame contact.
D. convection, conduction, and radiation.

_____ **48.** The concept of heat flow specifies that heat:
A. will flow from a hot substance to a cold substance.
B. will flow from a cold substance to a hot substance.
C. will vaporize a liquid thereby also reducing itself.
D. can travel through a building by one of four methods.

_____ **49.** In order to have spontaneous heating, the substance **must** be:
A. insulated.
B. organic.
C. inorganic.
D. exposed to sunlight.

_____ **50.** The two types of valves found in water supply distribution systems are:
A. gate and indicating.
B. ball and check.
C. screw and yoke.
D. indicating and non-indicating.

_____ **51.** The four fundamental components of a modern water system are:
A. source, mains, feeders, and risers.
B. primary, secondary, standpipes, and subscriber connections.
C. pipes, valves, hydrants, and pumps.
D. source, treatment plant, means of moving, and delivery system.

_____ **52.** In an assembly area of less concentrated use, such as a conference room or dining room measuring 150 feet by 200 feet, what would be the occupant load, if a load factor of 15 sq. ft. per person is used?
A. 1,000
B. 2,000
C. 500
D. 5,000

_____ **53.** The **primary** method used in determining the occupant load of a structure or part of a structure is the _________________ method.
A. floor area
B. building height
C. occupancy classification
D. building exit
E. aisle

_____ **54.** During an inspection of a building, the first area to be inspected should be the:
A. roof.
B. basement.
C. interior.
D. exterior.

_____ **55.** A self-closing fire door is one that:
A. closes when heat activates the closing device.
B. closes by magnetic controls.
C. after being opened, returns to the closed position.
D. is a Class B door only.

_____ **56.** In terms of life safety, two of the **most** **important** functions of doors are to:

A. provide light and fresh air in case of a fire.

B. provide access for the fire and police department in times of an emergency.

C. help with the ventilation and confinement of the fire.

D. act as a barrier to fire and smoke and serve as components in a means of egress.

_____ **57.** That portion of a means of egress between the termination of an exit and a public way is an exit:

A. passage.

B. path.

C. access.

D. discharge.

_____ **58.** The **minimum** number of exits required from any story or portion thereof with an occupant load of 500 or less shall be:

A. 4.

B. 2.

C. 5.

D. 3.

_____ **59.** _________________ fixed fire extinguishing systems are similar to halon systems in that they are **primarily** used to protect against fires involving flammable gases or liquids and in electrical equipment.

A. Chemical

B. Carbon Dioxide

C. Foam

D. Automatic Sprinkler

_____ **60.** The standard for the installation of sprinkler systems in one- and two-family dwellings and manufactured homes is:

A. NFPA 20.

B. NFPA 13.

C. NFPA 25.

D. NFPA 13R.

E. NFPA 13D.

_____ **61.** The standard for the installation of sprinkler systems in residential occupancies up to and including four stories in height is:

A. NFPA 13D

B. NFPA 13R

C. NFPA 20

D. NFPA 13

E. NFPA 25

_____ **62.** NFPA recommends that fire hydrants flowing 1,500 gpm or greater be color coded:

A. red.

B. light blue.

C. green.

D. orange.

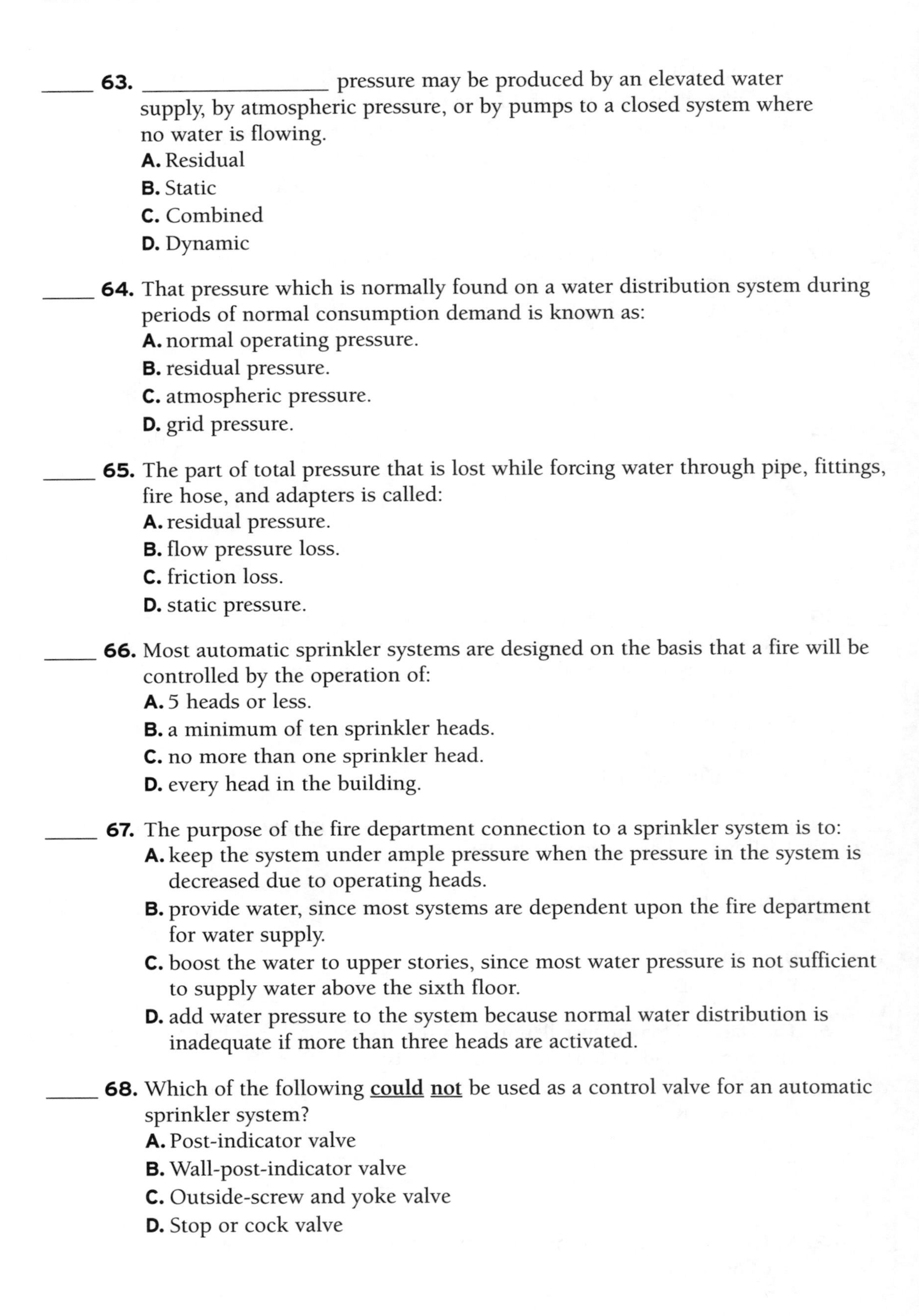

_____ **63.** ________________ pressure may be produced by an elevated water supply, by atmospheric pressure, or by pumps to a closed system where no water is flowing.

A. Residual
B. Static
C. Combined
D. Dynamic

_____ **64.** That pressure which is normally found on a water distribution system during periods of normal consumption demand is known as:

A. normal operating pressure.
B. residual pressure.
C. atmospheric pressure.
D. grid pressure.

_____ **65.** The part of total pressure that is lost while forcing water through pipe, fittings, fire hose, and adapters is called:

A. residual pressure.
B. flow pressure loss.
C. friction loss.
D. static pressure.

_____ **66.** Most automatic sprinkler systems are designed on the basis that a fire will be controlled by the operation of:

A. 5 heads or less.
B. a minimum of ten sprinkler heads.
C. no more than one sprinkler head.
D. every head in the building.

_____ **67.** The purpose of the fire department connection to a sprinkler system is to:

A. keep the system under ample pressure when the pressure in the system is decreased due to operating heads.
B. provide water, since most systems are dependent upon the fire department for water supply.
C. boost the water to upper stories, since most water pressure is not sufficient to supply water above the sixth floor.
D. add water pressure to the system because normal water distribution is inadequate if more than three heads are activated.

_____ **68.** Which of the following could not be used as a control valve for an automatic sprinkler system?

A. Post-indicator valve
B. Wall-post-indicator valve
C. Outside-screw and yoke valve
D. Stop or cock valve

_____ **69.** On an average, about _______________ percent of all fires in buildings with sprinklers are either extinguished or controlled by the sprinklers until they are completely extinguished by firefighters.
A. 25
B. 65
C. 85
D. 96

_____ **70.** _______________ sprinkler systems should be used in buildings where insufficient heat is maintained.
A. Wet pipe
B. Dry pipe
C. Deluge
D. Antifreeze

_____ **71.** The common classifications of sprinkler systems include all of the following **except**:
A. wet-pipe system.
B. dry-pipe system.
C. deluge system.
D. external-supply system.

_____ **72.** **Directions**: Place the following steps in the order necessary to perform a main drain test on a wet pipe sprinkler system.

1. Fully open the two-inch main drain, observe and record the pressure drop.
2. Close the two-inch main drain and compare readings to previously recorded readings.
3. Observe and record the pressure on the gauge(s) at the system riser.

A. 3,1,2
B. 2,3,1
C. 1,2,3
D. 3,2,1

_____ **73.** The _______________ system is ordinarily equipped with all sprinkler heads of the open type.
A. wet-pipe
B. dry-pipe
C. deluge
D. pre-action

_____ **74.** The full name of an OS&Y valve is:
A. open-system and yard valve.
B. open, shut, and yield valve.
C. outside-shield and yard valve.
D. outside-screw and yoke valve.

_____ **75.** Every sprinkler system should be equipped with a main control valve located between the:
A. riser and the branches.
B. source of water supply and the sprinkler system.
C. cross mains and the riser.
D. fire department connection and the riser.

_____ **76.** The control valve for a sprinkler system may be located at the system or outside the building. This valve should always be a/an:
A. check valve.
B. indicating type valve.
C. non-indicating type valve.
D. quarter turn valve.

_____ **77.** An OS&Y valve is open when the threads are:
A. retracted.
B. extended.
C. reversed.
D. crossed.

_____ **78.** A riser is a ________________ pipe which supplies the sprinkler system.
A. lateral
B. horizontal
C. diagonal
D. vertical

_____ **79.** In a pipe schedule designed sprinkler system, the water supply must be capable of delivering the required volume of water to the highest sprinkler head in the building at a ________________ pressure of ________________ psi.
A. static, 7
B. static, 15
C. residual, 7
D. residual, 15

_____ **80.** The trouble signal of a fire alarm system is dependent on the:
A. signaling circuit.
B. initiating devices.
C. indicating devices.
D. supervisory circuit.

_____ **81.** Which detector is <u>**most**</u> apt to give a false fire indication alarm when monitoring an area where arc welding is taking place?
A. Photoelectric
B. Ionization
C. Ultraviolet
D. Infrared

_____ **82.** Energized electrical equipment and the surrounding area are **best** protected with extinguishers that have a _______________ rating.
A. Class A
B. Class B
C. Class C
D. Class D

_____ **83.** Extinguishers suitable for Class A fires should be identified by a _______________ containing the letter "A." If colored, it should be _______________.
A. circle, blue
B. star, yellow
C. triangle, green
D. square, red

_____ **84.** A red square containing a letter would indicate an extinguisher may be used on _______________ fires.
A. Class A
B. Class B
C. Class C
D. Class D

_____ **85.** Fires involving flammable liquids, greases, and gases are _______________ fires.
A. Class A
B. Class B
C. Class C
D. Class D

_____ **86.** When approaching a fire, stored pressure water extinguishers are designed to be carried in a/an _______________ position and the operator is cautioned to keep the _______________ in hand at all times.
A. inverted, nozzle
B. upright, hose
C. horizontal, release grip
D. vertical, pin

_____ **87.** A blue circle with a letter designation in the center would indicate an extinguisher is rated for use on _______________ fires.
A. Class A
B. Class B
C. Class C
D. Class D

_____ **88.** When considering the distribution of fire extinguishers for Class A hazards, the **maximum** travel distance to extinguishers will be _______________ feet.
A. 50
B. 75
C. 150
D. 30

_____ **89.** A stored-pressure water extinguisher would be chosen to attack a _______________ fire.
A. Class A
B. Class B
C. Class C
D. Class D

_____ **90.** The preferred method of applying a portable extinguishing agent to a fire is to direct the stream:
A. into the flame and allow it to settle.
B. a minimum of 5 feet in front of the spill to prevent agitation.
C. up-wind and allow it to be blown on to the fire.
D. at the base of the fire using a sweeping motion.

_____ **91.** Carbon dioxide extinguishers are generally rated for Class _______________ fires.
A. A and B
B. B and C
C. C and D
D. B and D

_____ **92.** A multipurpose dry-chemical fire extinguisher **will not be** effective on which of the following classes of fires?
A. Flammable liquids
B. Ordinary combustibles
C. Electrical fires
D. Combustible metals

_____ **93.** For all types of kitchen cooking equipment, a common fire hazard is the potential for ignition of:
A. gas vapors.
B. electrical connectors.
C. grease filters.
D. adjacent combustible materials.

_____ **94.** Storage of combustible waste within a structure is known as a/an:
A. fire hazard.
B. ignition source.
C. target hazard.
D. heat-source hazard.

_____ **95.** A hazard that arises from processes or operations that are related to a specific occupancy defines _______________ hazard.
A. special
B. common
C. expected
D. target

_____ **96.** The flow of electricity through a wiring system is similar to the flow of water through a water distribution system. Which of the following **best** describes this relationship?

A. Quantity = GPM/Amperage
Pressure = PSI/Voltage
Resistance = Friction Loss/OHMS

B. Pressure = PSI/Amperage
Resistance = Friction Loss/OHMS
Quantity = GPM/Voltage

C. Resistance = Friction Loss/Amperage
Quantity = GPM/OHMS
Pressure = PSI/Voltage

D. Quantity = GPM/OHMS
Pressure = PSI/Voltage
Quantity = Friction Loss/Amperage

_____ **97.** The terms bonding and grounding refer to the process of:

A. an increase in the electrical potential between two wires.
B. dissipating static charges.
C. the flow of protons through a wire.
D. connecting a static generator to a nonconductive object.

_____ **98.** Fire spread through a ventilation duct system is generally controlled by:

A. fire stops.
B. smoke detector activation.
C. duct of noncombustible construction.
D. fire dampers.

_____ **99.** Fire inspectors may want to acquaint fire suppression forces with proposed building plans so they can:

A. recommend code violations.
B. point out better building construction methods.
C. provide their signatures on the plan review process.
D. provide feedback about problems that may be encountered during a fire.

_____ **100.** A floor plan sketch or diagram of a structure consists of:

A. a detailed view of the construction features of each floor of a building.
B. an outline of each floor of the building showing walls, partitions, openings, etc.
C. a cutaway view of a particular portion of a building.
D. showing the building and grounds as they are actually seen in depth by the eye.

Now that you have finished the feedback step for Examination I-1, it is time to repeat the process by taking another comprehensive examination of NFPA Standard 1031.

Did you score higher than 80 percent on Examination I-1? Circle Yes or No in ink. (We will return to your answer to this question later in SAEP.)

Examination I-2: Adding Difficulty and Depth

During Examination I-2, progress will be made in developing depth of knowledge and skills.

Step 1—Take Examination I-2. When you have completed Examination I-2, compare your answers with the correct answers in Appendix A.

Step 2—Score Examination I-2. How many examination items did you miss? Write the number of missed examination items in the blank in ink ________________. Enter the number of examination items you guessed in this blank ________________. Enter these numbers in the designated locations in your Personal Progress Plotter.

Step 3—Once again, the learning begins. During the feedback step, research the correct answer using Appendix A information for Examination I-2. Highlight the correct answer during your research of the reference materials. Read the entire paragraph containing the correct answer.

Helpful Hint

Follow each step carefully to realize the best return on effort. Would you consider investing your money in a venture without some chance of return on that investment? Examination preparation is no different. You are investing time and expecting a significant return for that time. If, indeed, time is money, then you are investing money and are due a return on that investment. Doing things right and doing the right things in examination preparation will ensure the maximum return on effort.

Examination I-2

Directions

Remove Examination I-2 from the manual. First, take a careful look at the examination. There should be 100 examination items. Notice that a blank line precedes each examination item number. Enter the answer to the examination item on this line. Write the answer in ink. Remember the rule about not changing your answers. Changed answers are most often changed to an incorrect answer, and more often than not the answer that is chosen first is correct.

If you guess an answer, place an "X" or a check mark by your answer. This step is vitally important to gain and master knowledge. We will explain how we treat the "guessed" items later in SAEP.

Take the examination. Once you complete it, go to Appendix A and score your examination. Carefully follow the directions for feedback of the missed and guessed examination items.

_____ **1.** Written letters and reports serve not only as records of an inspection, but also can be used:

A. to coerce the owner to not seek an appeal.
B. as a basis for legal action.
C. to enforce hazardous materials storage requirements.
D. as leverage against the manager of the property.

_____ **2.** **Directions**: Select the correct answer from the list below, then choose the answer from A-E which corresponds to your selection.

The item on the following list that need **not** be recorded on an inspection report is:

1. the name and address of the business.
2. the date of inspection.
3. the names of inspector(s), business or property owner(s), and individual with the inspector.
4. a list of violation(s) and applicable code reference.
5. detailed building plans.

A. 3
B. 5
C. 1
D. 2
E. 4

_____ **3.** An inspection report would be of little value if it **did not** contain the:

A. home address and phone number of the inspector.
B. complete floor plans.
C. name and address of the business.
D. department's ability to handle a fire.

_____ **4.** In an inspection report, fire inspectors are generally concerned with presenting the facts and evidence to prove a point, draw a conclusion, or:
A. justify an opinion.
B. show personal concern.
C. justify a recommendation.
D. reflect on a problem.

_____ **5.** Most local agencies have ordinances that provide for issuance of permits in order to:
A. generate records for tax appraisal needs of the county.
B. influence contractors to provide building services.
C. monitor special operations and conditions within the jurisdiction.
D. guide the city council in establishing the local tax levy rates.

_____ **6.** The following are all elements of what type of plan?

- The location of hydrants and water mains
- The slopes or grades affecting placement of ground ladders
- Grades on roadways affecting placement of fire department apparatus
- Actual location of fire department connections

A. Elevated view
B. Plan view
C. Floor
D. Site
E. Distribution

_____ **7.** The process of reviewing building plans and specifications to determine the safety characteristics of a proposed building refers to:
A. inspections.
B. citations.
C. plans review.
D. occupancy.

_____ **8.** When investigating a complaint, the inspector should be prepared to:
A. be told that no one is available to assist and be asked to come back later.
B. deal with possible negative attitudes of owners.
C. have difficulty with the receptionist.
D. be directed away from the area in question.

_____ **9.** Performance-based codes are difficult to enforce because:
A. performance of each inspector may be different.
B. performance of structural members is dependent upon supporting technical information of engineers and architects.
C. if data are entered inconsistently, performance will be inconsistent.
D. the code may change frequently, based on its performance.

_____ **10.** Under what circumstances might a fire inspector end up in court following an inspection?
A. The business owner felt he/she was treated unfairly.
B. The owner disagreed with the inspector's findings.
C. The property owner exhausted all other means of appeal and still is not satisfied with the outcome.
D. The property owner did not immediately correct the violations.

_____ **11.** When taking cases to court, the inspector may be placed in the position of fulfilling two roles, that of a(n) ________________ and a(n) ________________.
A. advisor, witness
B. witness, attorney
C. witness, fact recorder
D. advisor, defendant

_____ **12.** When acting as an advisor during a court case, which of the following **would not** be an inspector's area of responsibility?
A. Fire ordinances
B. Criminal law
C. Technical terms
D. Facts pertaining to the case

_____ **13.** According to the NFPA 101, Life Safety Code, a building, or portion of a building, able to accommodate fewer than 50 persons and used for educational purposes above the 12th grade is classified as a/an:
A. educational occupancy.
B. business occupancy.
C. institutional occupancy.
D. mercantile occupancy.
E. assembly occupancy.

_____ **14.** According to NFPA 101, Life Safety Code, a building or portion of a building used for gathering together 50 or more persons for such purposes as worship, entertainment, or dining is classified as a/an:
A. place of assembly.
B. business occupancy.
C. occupancy of unusual structures.
D. mercantile occupancy.
E. educational occupancy.

_____ **15.** During a school fire drill, teachers and staff should have responsibility for:
A. placing students' papers and reports in a safe place.
B. checking rest rooms or other rooms for students.
C. resetting fire alarm panels.
D. being sure classroom doors are locked.

_____ **16.** Fire drills in educational occupancies **should not be** conducted during:
A. classes.
B. gym or recess periods.
C. school assemblies.
D. the same time period on the same day of the week.

_____ **17.** Fire department access is determined by local codes or ordinances to specify:
A. width of roads.
B. angle of roads.
C. height of roads.
D. flatness of roads.

_____ **18.** Which container should a fire inspector accept for outside storage of flammable liquids?
A. Polychloride drum
B. Steel pail
C. Any container approved for the liquid
D. No regulations if stored outside

_____ **19.** All portable tanks greater than _________________ should be treated according to the same standards as fixed tanks.
A. 60 gallons
B. 120 gallons
C. 240 gallons
D. 660 gallons

_____ **20.** What size above-ground storage tank placed inside of a building requires it to be placed in a fire resistant room or enclosure?
A. 220 gallons or larger
B. 275 gallons or larger
C. 500 gallons or larger
D. 660 gallons or larger

_____ **21.** Vents on above-ground storage that are located in proximity to buildings or public places shall be _________________ feet from the ground, so that vapors are released outside and in a safe area.
A. 25
B. 15
C. 12
D. 7

_____ **22.** Vent pipes must enter underground storage tanks through the:
A. side of the tank and must be sloped so that they drain toward the tanks.
B. top of the tank and must be sloped so that they drain toward the tanks.
C. side of the tank and must be sloped so that they drain away from the tanks.
D. top of the tank and must be sloped so that they drain away from the tanks.

_____ **23.** How many feet away shall filling, discharge, and vapor recovery connections for underground storage tanks that are made and broken for Class I, II, and III-A liquids be from the nearest building openings?

A. 5 feet
B. 10 feet
C. 20 feet
D. 25 feet

_____ **24.** Which is a hazard associated with the handling and use of flammable and combustible liquids?

A. Ventilation
B. Static electricity
C. Closed containers
D. Bonding containers

_____ **25.** Loading and unloading stations for Class I liquids should be no closer than _______________ feet from storage tanks, property lines, or adjacent buildings?

A. 25
B. 10
C. 15
D. 50

_____ **26.** Which condition would be conducive to creating an explosive atmosphere in the storage, handling, and use of flammable and combustible liquids?

A. Inadequate lighting
B. Penetrated fire wall
C. Concentration of vapor
D. Open atmosphere

_____ **27.** The vents used for underground storage tanks should have at least a _______________-inch nominal inside diameter.

A. 2
B. ½
C. ⅓
D. 1

_____ **28.** Fire inspectors must ensure that measures are taken to prevent flammable or combustible liquids from accumulating under L.P. gas containers to reduce the risk of:

A. blow-off.
B. BLUEE.
C. BLEVE.
D. BELCH.

_____ **29.** The vent pipe from underground storage tanks storing Class I flammable liquids must be at least _________________ feet above the adjacent ground level.
A. 5
B. 20
C. 12
D. 8

_____ **30.** Buildings and warehouses storing flammable or combustible liquid containers must provide access aisles **at least** _________________ feet wide.
A. three
B. six
C. four
D. twelve

_____ **31.** The **maximum** allowable size for approved safety cans for flammable or combustible liquids is _________________ gallons.
A. 3
B. 2
C. 1
D. 5

_____ **32.** When transferring flammable or combustible liquids, what precaution must be taken to prevent a spark due to static electricity?
A. Wet the floor with water.
B. Insulate the containers.
C. Provide a bonding wire between the containers.
D. Use only a non-metallic transfer hose.

_____ **33.** Cryogenic liquids are stored in special containers. These containers are typically:
A. only double-walled construction.
B. single-walled container that is insulated.
C. double-walled containers of corrosion–resistant metal.
D. single-walled containers that are insulated and constructed with corrosion-resistant metal.

_____ **34.** Combustible dusts present a unique hazard because of their ability to form _________________ mixtures within an enclosure.
A. flammable
B. combustible
C. explosive
D. nonflammable

_____ **35.** Operations generating combustible dusts require several preventive measures. An important way to reduce the possibility of a dust explosion is to:
A. employ frequent housekeeping practices.
B. use a water wash system on all equipment.
C. install explosion vents or release devices.
D. post warning signs on all equipment.

_____ **36.** The phenomenon by which heat, smoke, and fire gases will travel upward to the highest point and become trapped, bank down, and spread out laterally is known as:

A. backdraft.

B. thermal layering.

C. flashover.

D. buoyancy.

_____ **37.** In tall buildings, layering on top floors may be delayed until there is:

A. sufficient heat buildup to cause the upward movement of smoke and fire gases.

B. admission of sufficient oxygen to cause flashover.

C. an adequate roof opening to discharge smoke and fire gases.

D. enough water applied to cool the gases as they are leaving the seat of the fire.

_____ **38.** Heat that results from the effects of either high-frequency-alternating current on a nonconductor or pulsating-direct current **best** defines _______________ heating.

A. resistance

B. dielectric

C. induction

D. static-electricity

_____ **39.** The gas which is a product of incomplete combustion is:

A. carbon monoxide.

B. carbon dioxide.

C. carbon.

D. PVC.

_____ **40.** Which of the following is considered to be a hazardous atmosphere encountered during fires?

A. Oxygen deficiency

B. Elevated temperatures

C. Smoke

D. All of the above are correct.

_____ **41.** In the tetrahedron concept, which of the following is called the reducing agent?

A. Fuel

B. Heat

C. Oxygen

D. Chemical chain reaction

_____ **42.** The third phase of burning is known as the:

A. incipient phase.

B. flame-producing phase.

C. hot smoldering phase.

D. free-burning phase.

_____ **43.** The transfer of heat by the movement of air or liquid would be considered:
A. conduction.
B. radiation.
C. convection.
D. direct flame contact.

_____ **44.** In examining the phases of fire, we find that a steady-state burning fire will have a ceiling temperature of **approximately**:
A. 900°F.
B. 1000°F.
C. 1300°F.
D. 1700°F.

_____ **45.** Which phase of fire is considered the period of free burning?
A. Incipient
B. Steady-state
C. Smoldering
D. Rapid-oxidation

_____ **46.** Which represents the fourth side of the geometric figure that resembles a pyramid and has taken the place of the fire triangle?
A. Atoms and radicals
B. New compounds
C. Chemical chain reaction
D. Flame interface

_____ **47.** The fourth component of the burning process is a _________________ reaction.
A. mechanical
B. chemical
C. endothermic
D. replenishment

_____ **48.** Which of the following is the **primary** element that makes up approximately 80 percent of the mixture of gases that we breathe?
A. Oxygen
B. Nitrogen
C. Carbon dioxide
D. Helium

_____ **49.** When a basement fire extends to the attic and roof area without involving the intervening floors, it is **most likely** the result of:
A. conduction.
B. convection.
C. radiation.
D. ventilation.

_____ **50.** Convection is:
A. transfer of heat through space by infrared rays.
B. transfer of heat through a solid medium.
C. not considered a method of heat transfer.
D. transfer of heat through liquids and gases by circulating currents.

_____ **51.** Which of the following <u>is</u> <u>not</u> one of the component parts of a dry-barrel fire hydrant?
A. Operating stem
B. Stem nut
C. Automatic check
D. Drain hole

_____ **52.** The following depicts a _______________ hydrant.

A. dry-barrel
B. wet-barrel
C. drafting
D. dry

_____ **53.** The occupant load for a building or room should be established:
A. only when there is a change in the occupancy classification.
B. during the plans review process.
C. during the highest occupancies of the area.
D. when construction on the building nears completion.

_____ **54.** The <u>first</u> step to take when calculating occupant load for a nightclub is to:
A. consult the requirements of the local code.
B. determine the square footage of the floor area.
C. divide the net floor area into the total square footage.
D. determine the number of exits and the width of each.

_____ **55.** Which of the following <u>is</u> <u>not</u> a component of a means of egress?
A. Exit access
B. Exit passageways
C. Exit
D. Exit discharge

_____ **56.** Which of the following **is not** allowed to be used as a part of a means of egress in new construction?
A. Ramps
B. Escalators
C. Exit passageways
D. Stairs

_____ **57.** Exterior walls and structural members that are of approved noncombustible or limited-combustible materials and interior structural members constructed wholly or partly of wood are considered ________________ construction.
A. Type I
B. Type II
C. Type III
D. Type IV

_____ **58.** The upright wood supporting members in a frame dwelling are called:
A. rafters.
B. sheathing.
C. joists.
D. studs.

_____ **59.** A fire wall must be continuous from the foundation floor to a height:
A. equal with the roof rafter.
B. equal or above the roof as required by the local building code.
C. six inches below the roof line.
D. at least equal with the attic floor.

_____ **60.** Fire stops are **generally**:
A. 2" × 4" pieces of wood placed horizontally in walls between studs.
B. 2" × 4"s placed vertically between studs.
C. boards or curtains hanging from the ceilings to confine the fire.
D. ventilation devices in the roof of buildings.

_____ **61.** All sprinkler system control valves must be of ________________ type.
A. butterfly
B. stop-cock
C. indicating
D. non-indicating

_____ **62.** When a fire hydrant receives water from two or more directions, it is said to have ________________ feed.
A. distributed
B. circulating
C. primary
D. secondary

_____ **63.** The following illustration depicts a _______________ hydrant.

A. dry-barrel
B. wet-barrel
C. drafting
D. dry

_____ **64.** Direct pumping water systems are those in which water:
A. moves directly into the distribution system by gravity flow.
B. is supplied directly into the distribution system from elevated storage tanks.
C. is pumped directly into the distribution system with no elevated storage.
D. is pumped through the distribution system back into the main water supply.

_____ **65.** Valve spacing for water systems in high value districts should be no more than _______________ feet.
A. 300
B. 500
C. 750
D. 1,000

_____ **66.** In a water supply system, the sizes of the water mains from the largest to the smallest are:
A. primary, distributor, secondary.
B. distributor, secondary, primary.
C. secondary, primary, distributor.
D. primary, secondary, distributor.

_____ **67.** The term "grid" is sometimes used to describe a network of water mains that make up a distribution system. Which of the following violates the principle of a grid system?
A. Primary feeders
B. Secondary feeders
C. Inter-connecting distributors
D. Dead-end water mains

_____ **68.** The required residual pressure when computing available water supply in a specific area is _______________ psi.
A. 5
B. 10
C. 15
D. 20

_____ **69.** Residual pressure is defined as:
A. stored potential energy available.
B. the pressure remaining in a water supply while water is flowing.
C. forward velocity pressure at the point of discharge.
D. the minimum pressure required in a residential area.

_____ **70.** The forward velocity pressure at a discharge opening that is recorded by a pitot tube and gauge is known as _______________ pressure.
A. flow
B. static
C. normal-operating
D. atmospheric

_____ **71.** NFPA recommends a fire hydrant flowing 1250 gpm be color coded:
A. red.
B. blue.
C. green.
D. orange.

_____ **72.** Which of the following **is not** one of the component parts of a dry-barrel fire hydrant?
A. Operating stem
B. Stem nut
C. Automatic check
D. Drain hole

_____ **73.** The term "grid" is sometimes used to describe a network of water mains that make up a distribution system. Which of the following **is not** a component of a grid system?
A. Primary feeders
B. Secondary feeders
C. Distributors
D. Risers

_____ **74.** A pitot tube and gauge are used to:
A. measure gallonage.
B. measure pressure loss.
C. take a pressure reading.
D. take a flow reading.

_____ **75.** The pressure of a water distribution system when no water is flowing is called _______________ pressure.
A. static
B. constant
C. residual
D. normal operating

_____ **76.** The hose connections to the FDC (Fire Department Connection) must be of the _________________ type and equipped with standard caps.
A. right-handed
B. left-handed
C. female
D. male

_____ **77.** The standard hydrostatic test for all piping in wet pipe systems **is not less** than _________________ for _________________.
A. 50 psi, one hour
B. 200 psi, two hours
C. 150 psi, twenty minutes
D. 100 psi, one and one-half hours

_____ **78.** Dry-standpipe systems must be hydrostatically tested:
A. twice a year.
B. once a year.
C. every two years.
D. every five years.

_____ **79.** Which automatic sprinkler system normally has open sprinkler heads?
A. dry pipe
B. pre-action
C. deluge
D. wet

_____ **80.** Which sprinkler system has air in piping, deluge valve, fire detection devices, and closed sprinklers?
A. wet
B. dry pipe
C. pre-action
D. deluge

_____ **81.** The four types of standpipe systems are:
A. wet standpipe, dry standpipe, vented standpipe, and fire department supplied dry standpipe.
B. fire department supplied dry standpipe, wet standpipe, manually activated dry standpipe, and vented standpipe.
C. fire department supplied dry standpipe, automatic water supplied dry standpipe, manually activated dry standpipe, and wet standpipe.
D. manually activated dry standpipe, wet standpipe, automatic water supplied dry standpipe, and vented dry standpipe.

_____ **82.** A _________________ detector **generally** responds more quickly to smoldering fires.
A. photoelectric
B. ionization
C. fire gas
D. fixed temperature

_____ **83.** A/An _________________ detector is one which responds to rapid changes in temperature.
A. disk-thermostat
B. bimetallic
C. infrared
D. rate-of-rise

_____ **84.** An alarm system which **does not** retransmit an alarm to any agency or group away from the premises is called a:
A. zone noncoded.
B. central station.
C. proprietary alarm.
D. local alarm.

_____ **85.** What is the basic difference between a pump tank extinguisher and an air pressurized water extinguisher?
A. They both use water for extinguishment.
B. The air pressurized uses air pressure for extinguishment.
C. The air pressurized uses air pressure to expel the water.
D. There is no difference between the two.

_____ **86.** Extinguishers suitable for Class B fires should be identified by a _________________ containing the letter "B."
A. blue circle
B. green triangle
C. red square
D. yellow star

_____ **87.** Extinguishers suitable for Class C fires should be identified by a _________________ containing the letter "C."
A. yellow star
B. green triangle
C. red square
D. blue circle

_____ **88.** A multipurpose dry chemical extinguisher would be rated for _________________ fires.
A. Class A and B
B. Class B and C
C. Class A and C
D. Class A, B, and C

_____ **89.** A Class D fire involves:
A. combustible metals.
B. flammable liquids.
C. electrical equipment.
D. wood, paper, etc.

_____ **90.** Class B fires involve:
A. flammable liquids.
B. energized electrical equipment.
C. ordinary combustibles.
D. combustible metals.

_____ **91.** Extinguishers suitable for Class D fires should be identified by a _________________ containing the letter "D."
A. blue circle
B. yellow star
C. green triangle
D. red square

_____ **92.** Class A fires involve:
A. ordinary combustibles.
B. electrical equipment.
C. flammable liquids.
D. combustible metals.

_____ **93.** Fires involving combustible metals, such as magnesium, titanium, zirconium, sodium, and potassium are _________________ fires.
A. Class A
B. Class B
C. Class C
D. Class D

_____ **94.** Dry powder extinguishers are rated for use on Class _________________ fires.
A. A
B. B
C. C
D. D

_____ **95.** All portable extinguishers are rated according to their:
A. size.
B. cooling potential.
C. intended use.
D. conductivity.

_____ **96.** A common hazard associated with commercial cooking equipment is:
A. improper installation of ventilation system.
B. an accumulation of carbon monoxide due to a lack of make-up air.
C. improper use of cooking equipment.
D. build-up of grease due to lack of cleaning or maintenance.

_____ **97.** The standard source for technical information related to electricity and electrical code can be found in:
A. Electrical Inspector's Handbook.
B. NFPA 70.
C. NFPA 72.
D. NFPA 101.

_____ **98.** Which NFPA Standard refers to the installation of equipment for the removal of smoke and grease-laden vapors from commercial cooking equipment?
A. NFPA 101
B. NFPA 96
C. NFPA 13
D. NFPA 10
E. NFPA 1031

_____ **99.** Draft curtains are sometimes used in large areas of buildings that are not otherwise subdivided. The purpose of draft curtains is to:
A. limit the mushrooming effect of heat and smoke.
B. help with protection of roof structure.
C. deflect hose streams into hard to reach areas.
D. stop the spread of fire through an assembly.

_____ **100.** When there is no ________________, code enforcement becomes a reactive process beginning after a building is constructed and ready for occupancy.
A. preconstruction inspection
B. plans review process
C. architect's approval
D. prefire planning

Helpful Hint

Try to determine why you selected the wrong answer. Usually something influenced your selection. Focus on the difference between your wrong answer and the correct answer. Carefully read and study the entire paragraph containing the correct answer. Highlight the answer just as you did for Examination I-1.

Did you score higher than 80 percent on Examination I-2? Circle Yes or No in ink. (We will return to your answer to this question later in SAEP.)

Examination I-3: Confirming What You Mastered

During Examination I-3, progress will be made in reinforcing what you have learned and improving your examination-taking skills. This examination contains approximately 60 percent of the examination items you have already answered and several new examination items. Follow the steps carefully to realize the best return on effort.

Step 1—Take Examination I-3. When you have completed Examination I-3, compare your answers with the correct answers in Appendix A.

Step 2—Score Examination I-3. How many examination items did you miss? Write the number of missed examination items in the blank in ink ________________. Enter the number of examination items you guessed in this blank ________________. Enter these numbers in the designated locations in your Personal Progress Plotter.

Step 3—Complete the Feedback Step. Research the correct answer using Appendix A information for Examination I-3. Highlight the correct answer during your research of the reference materials. Read the entire paragraph containing the correct answer.

Examination I-3

Directions

Remove Examination I-3 from the manual. First, take a careful look at the examination. There should be 150 examination items. Notice that a blank line precedes each examination item number. Enter the answer to the examination item on this line. Write the answer in ink. Remember the rule about not changing your answers. Changed answers are most often changed to an incorrect answer, and more often than not the answer that is chosen first is correct.

If you guess the answer to a question, place an "X" or a check mark by your answer. This step is vitally important to gain and master knowledge. We will explain how we treat the "guessed" items later in SAEP.

Take the examination. Once you complete it, go to Appendix A and score your examination. Carefully follow the directions for feedback of the missed and guessed examination items.

_____ **1.** Procedures in various jurisdictions may differ due to differences in the code enforcement each has adopted. These procedures should be:

A. restricted to a single typed page.
B. designed to promote appeals.
C. in detailed, written form.
D. developed in a way that can be easily changed.
E. maintained under strict security.

_____ **2.** Directions: Select the correct answer from the list below, then choose the answer from A-E which corresponds to your selection.

The item on the following list that need not be recorded on an inspection report is:

1. the name and address of the business.
2. the date of inspection.
3. the names of inspector(s), business or property owner(s), and individual with the inspector.
4. a list of violation(s) and applicable code reference.
5. detailed building plans.

A. 3
B. 5
C. 1
D. 2
E. 4

_____ **3.** An inspection report would be of little value if it did not contain the:

A. home address and phone number of the inspector.
B. complete floor plans.
C. name and address of the business.
D. department's ability to handle a fire.

_____ **4.** Before making an inspection, you should review _______________ to become familiar with a particular site.
A. existing records
B. a street map
C. the fire reports
D. insurance records

_____ **5.** It is generally considered most reliable to catalog fire prevention files by:
A. business name.
B. occupancy classification.
C. occupancy name.
D. street address.

_____ **6.** The purpose(s) of the inspection form is to:
A. provide documentation that an inspection took place.
B. document any violations that have been found.
C. provide documentation that violations were corrected on return visits.
D. All of the above.

_____ **7.** _______________ are a way of staying aware of changes in use or hazardous conditions.
A. Surveys
B. Permits
C. Self-inspections
D. Fire reports

_____ **8.** When making application for a permit, additional documentation may be required for submittal. Which of the following **would** **not** be documentation required for a permit?
A. Material safety data sheets (MSDS)
B. Hazardous materials storage manifest
C. Shop drawings
D. Construction documents

_____ **9.** On a site plan, the north directional symbol **usually** points toward the:
A. magnetic North Pole.
B. right side of the page.
C. top of the page.
D. true North Pole.

_____ **10.** Construction drawings are used to indicate all of the following **except**:
A. egress systems.
B. building size.
C. occupant load.
D. fire suppression capabilities.
E. method of construction.

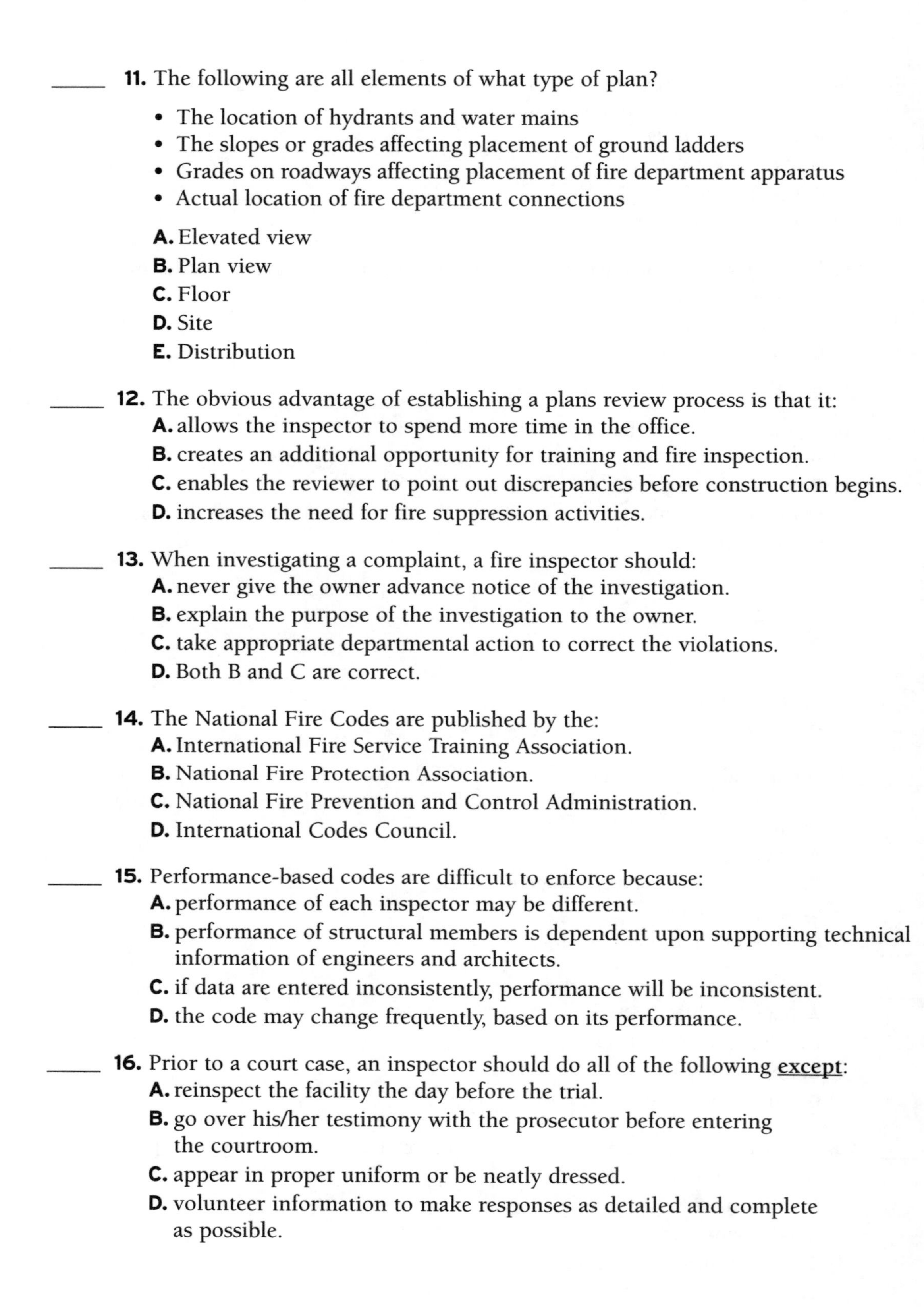

_____ **11.** The following are all elements of what type of plan?

- The location of hydrants and water mains
- The slopes or grades affecting placement of ground ladders
- Grades on roadways affecting placement of fire department apparatus
- Actual location of fire department connections

A. Elevated view
B. Plan view
C. Floor
D. Site
E. Distribution

_____ **12.** The obvious advantage of establishing a plans review process is that it:
A. allows the inspector to spend more time in the office.
B. creates an additional opportunity for training and fire inspection.
C. enables the reviewer to point out discrepancies before construction begins.
D. increases the need for fire suppression activities.

_____ **13.** When investigating a complaint, a fire inspector should:
A. never give the owner advance notice of the investigation.
B. explain the purpose of the investigation to the owner.
C. take appropriate departmental action to correct the violations.
D. Both B and C are correct.

_____ **14.** The National Fire Codes are published by the:
A. International Fire Service Training Association.
B. National Fire Protection Association.
C. National Fire Prevention and Control Administration.
D. International Codes Council.

_____ **15.** Performance-based codes are difficult to enforce because:
A. performance of each inspector may be different.
B. performance of structural members is dependent upon supporting technical information of engineers and architects.
C. if data are entered inconsistently, performance will be inconsistent.
D. the code may change frequently, based on its performance.

_____ **16.** Prior to a court case, an inspector should do all of the following **except**:
A. reinspect the facility the day before the trial.
B. go over his/her testimony with the prosecutor before entering the courtroom.
C. appear in proper uniform or be neatly dressed.
D. volunteer information to make responses as detailed and complete as possible.

_____ **17.** Which of the following <u>is</u> <u>not</u> suggested for courtroom procedure or behavior?
A. Never become argumentative on the witness stand.
B. Make sure that all physical evidence, exhibits, photographs, notes, and reference materials are brought to court.
C. Attempt to answer a question you do not know.
D. Remain impartial. Do not give the impression that you have a personal dislike for the defendant.

_____ **18.** Fire drills in educational occupancies <u>should</u> <u>not</u> <u>be</u> conducted:
A. during classroom changes, since the students are already in the hallways.
B. during assembly programs, since the students are all in one place making for easy evacuation.
C. during lunch time.
D. at the start of each day.

_____ **19.** Fire drills should be conducted in:
A. storage warehouses.
B. all occupancies.
C. schools and assembly areas only.
D. schools and homes only.

_____ **20.** Fire drills in educational occupancies <u>should</u> <u>not</u> <u>be</u> conducted during:
A. classes.
B. gym or recess periods.
C. school assemblies.
D. the same time period on the same day of the week.

_____ **21.** Private driveways or parking lots may consist of either a _______________ skin of asphalt over the top of gravel or a _______________ slab of concrete.
A. thin/thick
B. thin/thin
C. thick/thin
D. thick/thick

_____ **22.** Obstructions to fire department access include all the following <u>except</u>:
A. burglar bars.
B. ornamental walls.
C. false fronts.
D. ramps.

_____ **23.** The temperature at which a liquid fuel, once ignited, will continue to burn is known as:
A. fire point.
B. vapor temperature.
C. boiling point.
D. flash point.

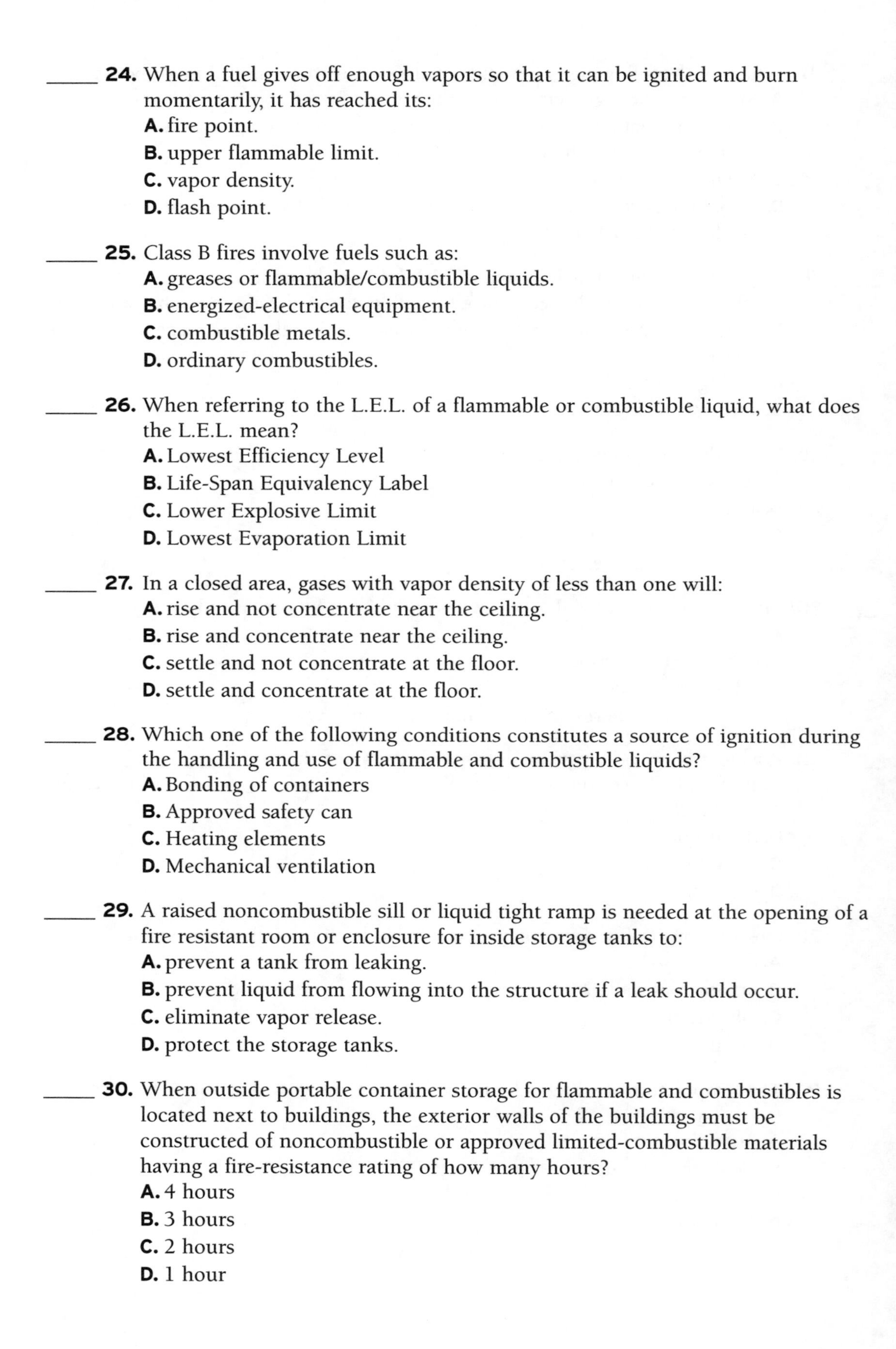

_____ **24.** When a fuel gives off enough vapors so that it can be ignited and burn momentarily, it has reached its:
A. fire point.
B. upper flammable limit.
C. vapor density.
D. flash point.

_____ **25.** Class B fires involve fuels such as:
A. greases or flammable/combustible liquids.
B. energized-electrical equipment.
C. combustible metals.
D. ordinary combustibles.

_____ **26.** When referring to the L.E.L. of a flammable or combustible liquid, what does the L.E.L. mean?
A. Lowest Efficiency Level
B. Life-Span Equivalency Label
C. Lower Explosive Limit
D. Lowest Evaporation Limit

_____ **27.** In a closed area, gases with vapor density of less than one will:
A. rise and not concentrate near the ceiling.
B. rise and concentrate near the ceiling.
C. settle and not concentrate at the floor.
D. settle and concentrate at the floor.

_____ **28.** Which one of the following conditions constitutes a source of ignition during the handling and use of flammable and combustible liquids?
A. Bonding of containers
B. Approved safety can
C. Heating elements
D. Mechanical ventilation

_____ **29.** A raised noncombustible sill or liquid tight ramp is needed at the opening of a fire resistant room or enclosure for inside storage tanks to:
A. prevent a tank from leaking.
B. prevent liquid from flowing into the structure if a leak should occur.
C. eliminate vapor release.
D. protect the storage tanks.

_____ **30.** When outside portable container storage for flammable and combustibles is located next to buildings, the exterior walls of the buildings must be constructed of noncombustible or approved limited-combustible materials having a fire-resistance rating of how many hours?
A. 4 hours
B. 3 hours
C. 2 hours
D. 1 hour

_____ **31.** Which container should a fire inspector accept for outside storage of flammable liquids?
A. Polychloride drum
B. Steel pail
C. Any container approved for the liquid
D. No regulations if stored outside

_____ **32.** All portable tanks greater than _______________ should be treated according to the same standards as fixed tanks.
A. 60 gallons
B. 120 gallons
C. 240 gallons
D. 660 gallons

_____ **33.** What size above-ground storage tank placed inside of a building requires it to be placed in a fire resistant room or enclosure?
A. 220 gallons or larger
B. 275 gallons or larger
C. 500 gallons or larger
D. 660 gallons or larger

_____ **34.** Which condition would be conducive to creating an explosive atmosphere in the storage, handling, and use of flammable and combustible liquids?
A. Inadequate lighting
B. Penetrated fire wall
C. Concentration of vapor
D. Open atmosphere

_____ **35.** Fire inspectors must ensure that measures are taken to prevent flammable or combustible liquids from accumulating under L.P. gas containers to reduce the risk of:
A. blow-off.
B. BLUEE.
C. BLEVE.
D. BELCH.

_____ **36.** The agency that regulates the transportation of compressed and liquefied gases is the:
A. U.S. Department of Transportation (DOT).
B. Petroleum Tank Institute (PTI).
C. National Transportation and Safety Board (NTSB).
D. Environmental Protection Agency (EPA).

_____ **37.** Compressed and liquefied gas cylinders should be stored in:
A. wet, cool locations.
B. designed rooms which are safe, dry, and well ventilated.
C. sheds with uneven fireproof floors.
D. outside open areas without protection from adverse weather conditions.

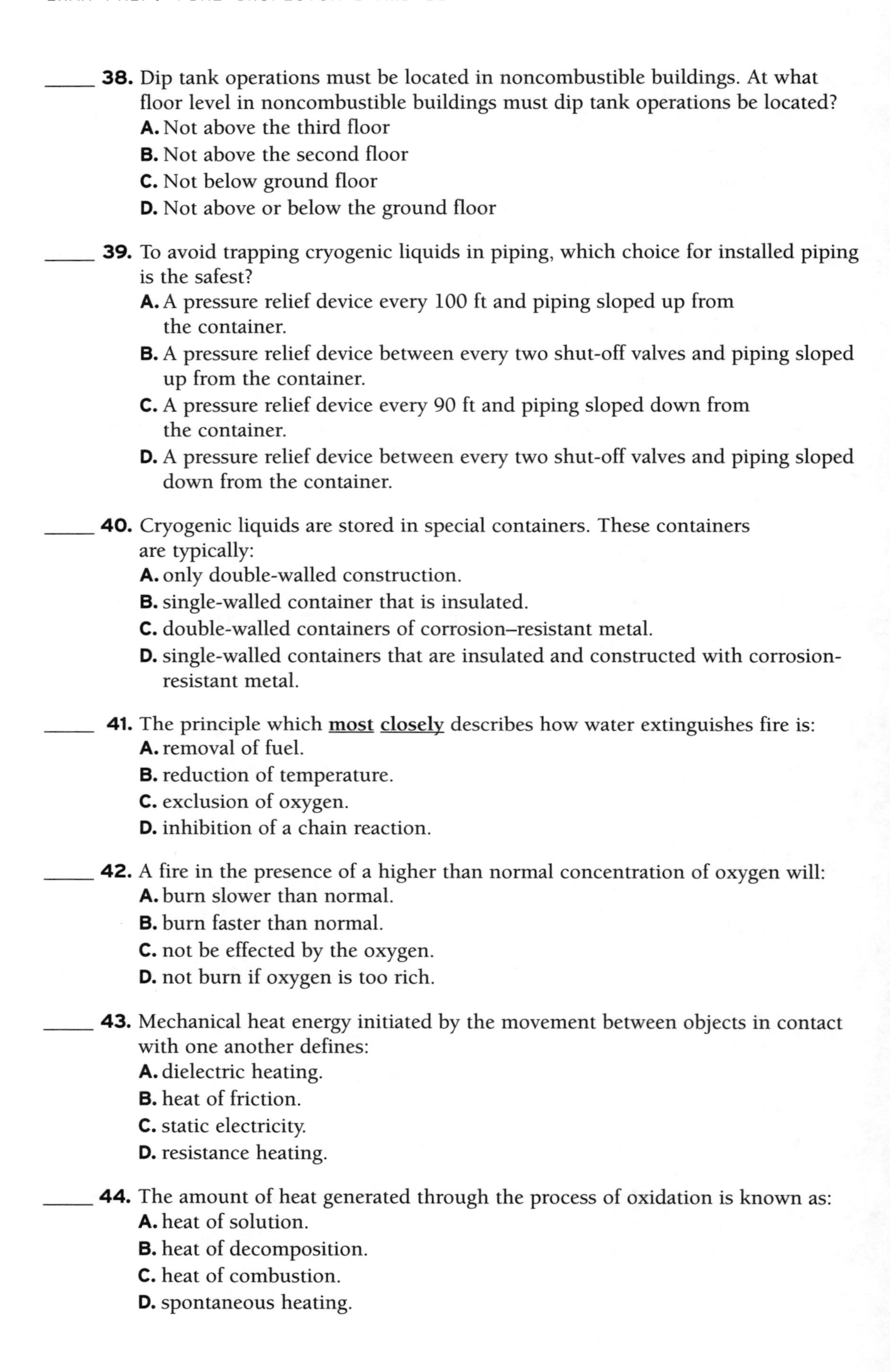

_____ **38.** Dip tank operations must be located in noncombustible buildings. At what floor level in noncombustible buildings must dip tank operations be located?

A. Not above the third floor
B. Not above the second floor
C. Not below ground floor
D. Not above or below the ground floor

_____ **39.** To avoid trapping cryogenic liquids in piping, which choice for installed piping is the safest?

A. A pressure relief device every 100 ft and piping sloped up from the container.
B. A pressure relief device between every two shut-off valves and piping sloped up from the container.
C. A pressure relief device every 90 ft and piping sloped down from the container.
D. A pressure relief device between every two shut-off valves and piping sloped down from the container.

_____ **40.** Cryogenic liquids are stored in special containers. These containers are typically:

A. only double-walled construction.
B. single-walled container that is insulated.
C. double-walled containers of corrosion–resistant metal.
D. single-walled containers that are insulated and constructed with corrosion-resistant metal.

_____ **41.** The principle which **most closely** describes how water extinguishes fire is:

A. removal of fuel.
B. reduction of temperature.
C. exclusion of oxygen.
D. inhibition of a chain reaction.

_____ **42.** A fire in the presence of a higher than normal concentration of oxygen will:

A. burn slower than normal.
B. burn faster than normal.
C. not be effected by the oxygen.
D. not burn if oxygen is too rich.

_____ **43.** Mechanical heat energy initiated by the movement between objects in contact with one another defines:

A. dielectric heating.
B. heat of friction.
C. static electricity.
D. resistance heating.

_____ **44.** The amount of heat generated through the process of oxidation is known as:

A. heat of solution.
B. heat of decomposition.
C. heat of combustion.
D. spontaneous heating.

_____ **45.** Heating of an organic substance without an external heat source is known as:
A. spontaneous heating.
B. heat of decomposition.
C. dielectric heating.
D. frictional heat.

_____ **46.** The release of heat usually due to bacterial action <u>**best**</u> defines:
A. spontaneous heating.
B. heat of decomposition.
C. heat of compression.
D. heat of solution.

_____ **47.** In order to have spontaneous heating, the substance must be:
A. insulated.
B. organic.
C. inorganic.
D. exposed to sunlight.

_____ **48.** The phenomenon by which heat, smoke, and fire gases will travel upward to the highest point and become trapped, bank down, and spread out laterally is known as:
A. backdraft.
B. thermal layering.
C. flashover.
D. buoyancy.

_____ **49.** In tall buildings, layering on top floors may be delayed until there is:
A. sufficient heat buildup to cause the upward movement of smoke and fire gases.
B. admission of sufficient oxygen to cause flashover.
C. An adequate roof opening to discharge smoke and fire gases.
D. Enough water applied to cool the gases as they are leaving the seat of the fire.

_____ **50.** Heat that results from the effects of either high-frequency-alternating current on a nonconductor or pulsating-direct current best defines _________________ heating.
A. resistance
B. dielectric
C. induction
D. static-electricity

_____ **51.** The gas which is a product of incomplete combustion is:
A. carbon monoxide.
B. carbon dioxide.
C. carbon.
D. PVC.

_____ **52.** In the tetrahedron concept, which of the following is called the reducing agent?
A. Fuel
B. Heat
C. Oxygen
D. Chemical chain reaction

_____ **53.** Which represents the fourth side of the geometric figure that resembles a pyramid and has taken the place of the fire triangle?
A. Atoms and radicals
B. New compounds
C. Chemical chain reaction
D. Flame interface

_____ **54.** The fourth component of the burning process is a _________________ reaction.
A. mechanical
B. chemical
C. endothermic
D. replenishment

_____ **55.** When a basement fire extends to the attic and roof area without involving the intervening floors, it is **most likely** the result of:
A. conduction.
B. convection.
C. radiation.
D. ventilation.

_____ **56.** In which of the following conditions is a solid fuel the most hazardous?
A. Powder or dusts
B. Turnings or ribbons
C. Shredded materials
D. All are equally dangerous.

_____ **57.** What kind of heat energy is the heat of compression?
A. Chemical
B. Electrical
C. Mechanical
D. Nuclear

_____ **58.** Conditions indicating a hot smoldering phase of fire would be:
A. temperature of the area of involvement over 1000°F.
B. ceiling temperature in excess of 1300°F.
C. rapid destruction of area and contents.
D. free burning with entire area in flame.

_____ **59.** Which of the following are examples of mechanical heat sources?
A. Heat of decomposition, heat of solution
B. Frictional heat, heat of compression
C. Induction heating, heat from arcing
D. Nuclear fission or fusion

_____ **60.** The term rollover is defined as that period in a fire when:
A. flames flash across the ceiling.
B. flames flash across the top of a flammable liquid.
C. the entire contents of a room or area is fully involved in fire.
D. flames roll across the ceiling of an area.

_____ **61.** A/An _______________ is defined as a condition that will encourage a fire to start or will increase the extent or severity of a fire.
A. fire hazard
B. ignition source
C. target hazard
D. common hazard

_____ **62.** A/An _______________ is defined as a condition which is prevalent in almost all occupancies and will encourage a fire to start.
A. fire hazard
B. ignition source
C. target hazard
D. common hazard

_____ **63.** Of the following categories of heat, which one would be a source of chemical heat energy?
A. Induction heating
B. Frictional heat
C. Spontaneous heating
D. Fusion heating

_____ **64.** The minimum temperature to which a fuel in a normal atmosphere must be heated in order to start a self-sustained combustion independent of an ignition source is known as:
A. oxidation process.
B. autoignition temperature.
C. fire/burning point.
D. flash point.

_____ **65.** In some areas, building and fire prevention codes require that the textiles used in theater scenery, as well as curtains and draperies used in public assembly occupancies, be treated with:
A. copper dioxide.
B. flame-retardant material.
C. Teflon spray.
D. borax-soap solution.

_____ **66.** The two types of valves found in water supply distribution systems are:
A. gate and indicating.
B. ball and check.
C. screw and yoke.
D. indicating and non-indicating.

_____ **67.** The four fundamental components of a modern water system are:
A. source, mains, feeders, and risers.
B. primary, secondary, standpipes, and subscriber connections.
C. pipes, valves, hydrants, and pumps.
D. source, treatment plant, means of moving, and delivery system.

_____ **68.** Which of the following **is not** one of the component parts of a dry-barrel fire hydrant?
A. Operating stem
B. Stem nut
C. Automatic check
D. Drain hole

_____ **69.** The following depicts a _________________ hydrant.

A. dry-barrel
B. wet-barrel
C. Drafting
D. dry

_____ **70.** The occupant load for a building or room should be established:
A. only when there is a change in the occupancy classification.
B. during the plans review process.
C. during the highest occupancies of the area.
D. when construction on the building nears completion.

_____ **71.** During an inspection of a building, the **first** area to be inspected should be the:
A. roof.
B. basement.
C. interior.
D. exterior.

_____ **72.** A self-closing fire door is one that:
A. closes when heat activates the closing device.
B. closes by magnetic controls.
C. after being opened, returns to the closed position.
D. is a Class B door only.

_____ **73.** In a new building, the **minimum** door opening in a means of egress is _______________ inches.

A. 36
B. 32
C. 28
D. 24

_____ **74.** A continuous and unobstructed way of exit travel from any point in a building or structure to a public way is a/an:

A. ramp.
B. means of egress.
C. corridor.
D. exit.

_____ **75.** The measurement for **maximum** travel distance starts from the ____________ portion of the occupancy, curving around any corners or obstructions with a ____________ foot clearance, and ending at the center of the exit doorway.

A. center, one
B. most remote, one
C. center, two
D. most remote, two

_____ **76.** Which of the following **is not** a component of a means of egress?

A. Exit access
B. Exit passageways
C. Exit
D. Exit discharge

_____ **77.** Which of the following **is not** allowed to be used as a part of a means of egress in new construction?

A. Ramps
B. Escalators
C. Exit passageways
D. Stairs

_____ **78.** Fire stops are generally:

A. 2" × 4" pieces of wood placed horizontally in walls between studs.
B. 2" × 4"s placed vertically between studs.
C. boards or curtains hanging from the ceilings to confine the fire.
D. ventilation devices in the roof of buildings.

_____ **79.** A fire stop serves:

A. to cut off draft within walls.
B. to route fire around hazard areas.
C. the same purpose as a water curtain.
D. the same purpose as a fire wall.

_____ **80.** The standard for the installation of sprinkler systems in one- and two-family dwellings and manufactured homes is:
A. NFPA 20.
B. NFPA 13.
C. NFPA 25.
D. NFPA 13R.
E. NFPA 13D.

_____ **81.** NFPA recommends that fire hydrants flowing 1,500 gpm or greater be color coded:
A. red.
B. light blue.
C. green.
D. orange.

_____ **82.** The part of total pressure that is lost while forcing water through pipe, fittings, fire hose, and adapters is called:
A. residual pressure.
B. flow pressure loss.
C. friction loss.
D. static pressure.

_____ **83.** Most automatic sprinkler systems are designed on the basis that a fire will be controlled by the operation of:
A. five heads or less.
B. a minimum of ten sprinkler heads.
C. no more than one sprinkler head.
D. every head in the building.

_____ **84.** The purpose of the fire department connection to a sprinkler system is to:
A. keep the system under ample pressure when the pressure in the system is decreased due to operating heads.
B. provide water, since most systems are dependent upon the fire department for water supply.
C. boost the water to upper stories, since most water pressure is not sufficient to supply water above the sixth floor.
D. add water pressure to the system because normal water distribution is inadequate if more than three heads are activated.

_____ **85.** On an average, about _________________ percent of all fires in buildings with sprinklers are either extinguished or controlled by the sprinklers until they are completely extinguished by firefighters.
A. 25
B. 65
C. 85
D. 96

_____ **86.** <u>Directions</u>: Place the following steps in the order necessary to perform a main drain test on a wet pipe sprinkler system:

1. Fully open the two-inch main drain, observe and record the pressure drop.
2. Close the two-inch main drain and compare readings to previously recorded readings.
3. Observe and record the pressure on the gauge(s) at the system riser.

A. 3,1,2
B. 2,3,1
C. 1,2,3
D. 3,2,1

_____ **87.** The _______________ system is ordinarily equipped with all sprinkler heads of the open type.
A. wet-pipe
B. dry-pipe
C. deluge
D. pre-action

_____ **88.** An OS&Y valve is open when the threads are:
A. retracted.
B. extended.
C. reversed.
D. crossed.

_____ **89.** A riser is a _______________ pipe which supplies the sprinkler system.
A. lateral
B. horizontal
C. diagonal
D. vertical

_____ **90.** When a fire hydrant receives water from two or more directions, it is said to have _______________ feed.
A. distributed
B. circulating
C. primary
D. secondary

_____ **91.** Direct pumping water systems are those in which water:
A. moves directly into the distribution system by gravity flow.
B. is supplied directly into the distribution system from elevated storage tanks.
C. is pumped directly into the distribution system with no elevated storage.
D. is pumped through the distribution system back into the main water supply.

_____ **92.** In a water supply system, the sizes of the water mains from the largest to the smallest are:
A. primary, distributor, secondary.
B. distributor, secondary, primary.
C. secondary, primary, distributor.
D. primary, secondary, distributor.

_____ **93.** The term "grid" is sometimes used to describe a network of water mains that make up a distribution system. Which of the following violates the principle of a grid system?
A. Primary feeders
B. Secondary feeders
C. Inter-connecting distributors
D. Dead-end water mains

_____ **94.** The required residual pressure when computing available water supply in a specific area is ________________ psi.
A. 5
B. 10
C. 15
D. 20

_____ **95.** Residual pressure is defined as:
A. stored potential energy available.
B. the pressure remaining in a water supply while water is flowing.
C. forward velocity pressure at the point of discharge.
D. the minimum pressure required in a residential area.

_____ **96.** Which of the following **is not** one of the component parts of a dry-barrel fire hydrant?
A. Operating stem
B. Stem nut
C. Automatic check
D. Drain hole

_____ **97.** Which NFPA standard deals with the installation of sprinkler systems?
A. NFPA 17
B. NFPA 13
C. NFPA 20
D. NFPA 10

_____ **98.** Dry-standpipe systems must be hydrostatically tested:
A. twice a year.
B. once a year.
C. every two years.
D. every five years.

_____ **99.** Which automatic sprinkler system normally has open sprinkler heads?
A. dry pipe
B. pre-action
C. deluge
D. wet

_____ **100.** Which sprinkler system has air in piping, deluge valve, fire detection devices, and closed sprinklers?
A. wet
B. dry pipe
C. pre-action
D. deluge

_____ **101.** Dry chemical fixed fire extinguishing systems <u>are</u> <u>not</u> recommended for use on:
A. flammable liquids.
B. flammable gases.
C. delicate electrical equipment.
D. paper storage.

_____ **102.** The four types of standpipe systems are:
A. wet standpipe, dry standpipe, vented standpipe, and fire department supplied dry standpipe.
B. fire department supplied dry standpipe, wet standpipe, manually activated dry standpipe, and vented standpipe.
C. fire department supplied dry standpipe, automatic water supplied dry standpipe, manually activated dry standpipe, and wet standpipe.
D. manually activated dry standpipe, wet standpipe, automatic water supplied dry standpipe, and vented dry standpipe.

_____ **103.** Class III standpipe systems are intended to be used by:
A. fire department personnel or building occupants.
B. only fire department personnel.
C. only trained staff.
D. only building occupants.

_____ **104.** The device utilized on a sprinkler system to prevent an accidental water-flow alarm due to pressure surges in the water supply is a/an:
A. exhauster.
B. multiplexing system.
C. retard chamber.
D. accelerator.

_____ **105.** Partitions, storage and all obstruction <u>should</u> <u>not</u> be within _______________ inches of sprinkler heads.
A. 6
B. 18
C. 12
D. 24

_____ **106.** After observing a trip test on a dry-pipe system, water and air pressure gauges should:
A. be the same as before the test.
B. indicate equal pressure.
C. not read greater than 110 psi.
D. return to zero.

_____ **107.** When inspecting any dry-pipe sprinkler system, the air pressure gauge should read **approximately**:
A. the same as the water pressure.
B. the same as recorded on the previous test.
C. three times atmospheric pressure.
D. half of the gauges full-scale pressure.

_____ **108.** Large pipes that carry large quantities of water to various points along the water supply system for distribution to smaller mains **best** defines:
A. primary feeders.
B. secondary feeders.
C. distributors.
D. grid network.

_____ **109.** The smaller internal grid arrangements of a water distribution system that feed hydrants as well as the domestic and commercial requirements **best** describes:
A. primary feeders.
B. secondary feeders.
C. distributors.
D. grid network.

_____ **110.** ________________ deposits consist of mud, clay, and dead organisms.
A. Rust
B. Sedimentary
C. Tuberculation
D. Encrustation

_____ **111.** When a water source **does not** have adequate elevation to create proper pressure for gravity flow, it is necessary to use:
A. larger size pipes in the mains.
B. pumps to raise the systems pressure.
C. computer controlled pressure regulators.
D. negative coefficient of friction loss.

_____ **112.** The full name of a PIV valve is:
A. position-indicator valve.
B. post-indicator valve.
C. plant-industrial valve.
D. positive-inspection valve.

_____ **113.** A fire department connection to a sprinkler system enables firefighters to:
A. connect hand lines for attacking the fire.
B. back up the first-attack units.
C. support system pressure.
D. test the system.

_____ **114.** The two basic types of fixed carbon dioxide extinguishing systems are local application and:
A. hand-held.
B. total flooding.
C. deflected stream.
D. rate-of-release.

_____ **115.** Which of the following **would not** be considered an indicating valve?
A. Check valve
B. Outside screw and yoke valve
C. WPIV
D. PIV

_____ **116.** The trouble signal of a fire alarm system is dependent on the:
A. signaling circuit.
B. initiating devices.
C. indicating devices.
D. supervisory circuit.

_____ **117.** An alarm system which **does not** retransmit an alarm to any agency or group away from the premises is called a:
A. zone noncoded.
B. central station.
C. proprietary alarm.
D. local alarm.

_____ **118.** A local protective signaling system is intended **primarily** to alert:
A. the fire department dispatcher.
B. a private alarm contractor.
C. occupants of the protected area.
D. central monitoring station.

_____ **119.** Smoke and fire dampers are typically located at points where:
A. the hood system covers fuel-powered cooking equipment.
B. smoke and fire incidents are predicted to occur in the structure.
C. smoke and fire conditions are intensified by flammable liquids.
D. HVAC ducts penetrate smoke and fire barrier walls.

_____ **120.** Energized electrical equipment and the surrounding area are **best** protected with extinguishers that have a ________________ rating.
A. Class A
B. Class B
C. Class C
D. Class D

_____ **121.** Fires involving flammable liquids, greases, and gases are _______________ fires.

A. Class A
B. Class B
C. Class C
D. Class D

_____ **122.** When considering the distribution of fire extinguishers for Class A hazards, the **maximum** travel distance to extinguishers will be _______________ feet.

A. 50
B. 75
C. 150
D. 30

_____ **123.** A stored-pressure water extinguisher would be chosen to attack a _______________ fire.

A. Class A
B. Class B
C. Class C
D. Class D

_____ **124.** Extinguishers suitable for Class C fires should be identified by a _______________ containing the letter "C."

A. yellow star
B. green triangle
C. red square
D. blue circle

_____ **125.** A multipurpose dry chemical extinguisher would be rated for _______________ fires.

A. Class A and B
B. Class B and C
C. Class A and C
D. Class A, B, and C

_____ **126.** Class B fires involve:

A. flammable liquids.
B. energized electrical equipment.
C. ordinary combustibles.
D. combustible metals.

_____ **127.** Extinguishers suitable for Class D fires should be identified by a _______________ containing the letter "D."

A. blue circle
B. yellow star
C. green triangle
D. red square

_____**128.** Fires involving combustible metals, such as magnesium, titanium, zirconium, sodium, and potassium are ________________ fires.

A. Class A
B. Class B
C. Class C
D. Class D

_____**129.** A carbon dioxide (CO_2) extinguisher's principle means of discharge is:

A. chemical reaction.
B. stored internal pressure.
C. cartridge activation.
D. manual hand-pump.

_____**130.** A fire extinguisher bearing the picture symbol illustrated below would be suitable for extinguishing:

A. Class A, B, C, and D fires.
B. Class A, B, and C fires.
C. Class A and B only.
D. Class D only.

_____**131.** A wet chemical fire extinguisher used for cooking fires is:

A. AFF.
B. Class BC.
C. Class K.
D. FFF.

_____**132.** Extinguishers weighing less than 40 pounds (18.1 Kg.) should be installed with the top <u>not</u> <u>more</u> than ________________ from the floor.

A. 3 feet (1.0 m)
B. 4 feet (1.2 m)
C. 5 feet (1.5 m)
D. 6 feet (1.8 m)

_____**133.** Portable fire extinguishers should be thoroughly inspected:

A. annually.
B. twice a year.
C. monthly.
D. four times a year.

_____**134.** All maintenance procedures should include a thorough examination of these three basic parts of an extinguisher:
A. mechanical parts, extinguishing agent, and expelling means.
B. mechanical parts, color, and expelling means.
C. gauge, size, and color.
D. extinguishing agent, nozzle, and gauge.

_____**135.** How much water is required for a fire extinguisher to receive a 2-A rating?
A. 2 gallons
B. 1-1/4 gallons
C. 2-1/2 gallons
D. 1-3/4 gallons

_____**136.** Fires involving magnesium, titanium, zirconium, sodium, and potassium are _________________ fires.
A. Class A
B. Class B
C. Class K
D. Class D

_____**137.** Which of the following types of portable fire extinguishers is the **least likely** to expose the operator to electrical shock?
A. Dry chemical
B. Wet chemical
C. Stored-pressure AFFF
D. FFFP foam

_____**138.** Which group listed below contains the **most commonly** encountered combustible metals?
A. Magnesium, sodium, potassium, calcium
B. Sulfur, sodium, steel, calcium
C. Magnesium, sodium, calcium, sulfur
D. Sulfur, steel, magnesium, calcium

_____**139.** The classification of portable fire extinguishers required for plastic materials is:
A. Class A.
B. Class B.
C. Class C.
D. Class D.

_____**140.** The classification of portable fire extinguishers required for fires involving flammable or combustible liquids, greases and gases is:
A. Class A.
B. Class B.
C. Class C.
D. Class D.

_____ **141.** Fire extinguishers shall be visually inspected by the occupant or building owner:
A. yearly.
B. quarterly.
C. monthly.
D. every other year.

_____ **142.** A hazard that arises from processes or operations that are related to a specific occupancy defines _______________ hazard.
A. special
B. common
C. expected
D. target

_____ **143.** The terms bonding and grounding refer to the process of:
A. an increase in the electrical potential between two wires.
B. dissipating static charges.
C. the flow of protons through a wire.
D. connecting a static generator to a nonconductive object.

_____ **144.** Electrical heat energy in the form of an arc between oppositely charged conductors best defines:
A. dielectric heating.
B. induction heating.
C. frictional heat.
D. static electricity heating.

_____ **145.** The common hazard associated with central heating appliances, unit heaters, and room heaters is:
A. the thermostat temperature limit controls.
B. proximity of installation to combustibles.
C. asphyxiation due to inadequate make-up air.
D. the explosion potential.

_____ **146.** A common hazard associated with commercial cooking equipment is:
A. improper installations of ventilation system.
B. an accumulation of carbon monoxide due to a lack of make-up air.
C. improper use of cooking equipment.
D. build-up of grease due to lack of cleaning or maintenance.

_____ **147.** Most common refrigerants used in air conditioning and ventilation systems have a classification level of _______________ and _______________, as determined by the American National Standards Institute (ANSI).
A. toxicity, flammability
B. flame spread, smoke development
C. flammability, reactivity
D. explosive, reactivity

_____**148.** One of the best ways to impress the importance of inspections upon building owners is through:
A. a follow-up visit.
B. surprise inspections.
C. imposition of a large fine.
D. reporting favorable findings.

_____**149.** Fire spread through a ventilation duct system is generally controlled by:
A. fire stops.
B. smoke detector activation.
C. duct of noncombustible construction.
D. fire dampers.

_____**150.** Fire inspectors may want to acquaint fire suppression forces with proposed building plans so they can:
A. recommend code violations.
B. point out better building construction methods.
C. provide their signatures on the plan review process.
D. provide feedback about problems that may be encountered during a fire.

Did you score higher than 80 percent on Examination I-3? Circle Yes or No in ink.

Feedback Step

What do we do with your "yes" and "no" answers through the SAEP process? First, return to any "no" response. Go back to the highlighted answers for those examination items missed, and read and study the paragraph preceding the location of the answer as well as the paragraph following the paragraph where the answer is located. This will expand your knowledge base for the missed question, put it in a broader context, and improve associative learning. Remember, we are trying to develop mastery of the required knowledge. Scoring 80 percent on an examination is good but it is not mastery performance. To come out in the top of your group, you must score much higher than 80 percent on your training, promotion, or certification examination.

Carefully review "Summary of Key Rules for Taking an Examination" and "Summary of Helpful Hints" on the next two pages. Do this review now and at least two additional times prior to taking your next examination.

Helpful Hint

Studying the correct answers for missed items is a critical step in return on effort! The focus of attention is broadened and new knowledge is often gained by expanding association and contextual learning. During our research and field examination, self-study during this step of SAEP resulted in gains of 17 points from the first examination administered to the third examination. An increase in your score of 17 points can move you from the lower middle to the top of the list of persons taking a training, promotion, or certification examination. This is a competitive edge and a prime example of return on effort in action. Remember: Maximum effort = maximum results!

Summary of Key Rules for Taking an Examination

Rule 1—Examination preparation is not easy. Preparation is 95 percent perspiration and 5 percent inspiration.

Rule 2—Follow the steps very carefully. Do not try to reinvent or shortcut the system. It really works just as it was designed to!

Rule 3—Mark with an "X" any examination items for which you guessed the answer. To obtain the maximum return on effort, research any answer that you guessed even if you guessed correctly. Find the correct answer, highlight it, and then read the entire paragraph that contains the answer. Be honest and mark all questions on which you guessed. Some examinations have a correction for guessing built into the scoring process. The correction for guessing can reduce your final examination score. If you are guessing, you are not mastering the material.

Rule 4—Read questions twice if you have any misunderstanding and especially if the question contains complex directions or activities.

Rule 5—If you want someone to perform effectively and efficiently on the job, the training and testing program must be aligned to achieve this result.

Rule 6—When preparing examination items for job-specific requirements, the writer must be a subject matter expert with current experience at the level that the technical information is applied.

Rule 7—Good luck = good preparation.

Summary of Helpful Hints

Helpful Hint—Your first impression is often the best. More than 41 percent of changed answers during our SAEP field test were changed from a right answer to a wrong answer. Another 33 percent changed from a wrong answer to another wrong answer. Only 26 percent of answers were changed from wrong to right. In fact, three participants did not make a perfect score of 100 percent because they changed one right answer to a wrong one! Think twice before you change your answer. The odds are not in your favor.

Helpful Hint—Researching correct answers is one of the most important activities in SAEP. Locate the correct answer for all missed examination items. Highlight the correct answer. Then read the entire paragraph containing the answer. This will put the answer in context for you and provide important learning by association.

Helpful Hint—Work through all missed examination items using the same technique. Reading the entire paragraph improves retention of the information and helps you develop an association with the material and learn the correct answers. This step may sound simple. A major finding during the development and field testing of SAEP was that you learn from your mistakes.

Helpful Hint—Follow the steps carefully to realize the best return on effort. Would you consider investing your money in a venture without some chance of return on that investment? Examination preparation is no different. You are investing time and expecting a significant return for that time. If, indeed, time is money, then you are investing money and are due a return on that investment.

Helpful Hint—Try to determine why you selected the wrong answer. Usually something influenced your selection. Focus on the difference between your wrong answer and the correct answer. Carefully read and study the entire paragraph containing the correct answer.

Helpful Hint—Studying the correct answers for missed items is a critical step in return on effort! The focus of attention is broadened and new knowledge is often gained by expanding association and contextual learning. During our research and field test, self-study during this step of SAEP resulted in gains of 17 points from the first examination administered to the third examination. An increase in your score of 17 points can move you from the lower middle to the top of the list of persons taking a training, promotion, or certification examination. That is a competitive edge and a prime example of return on effort in action. Remember: Maximum effort = maximum results!

PHASE II

Fire Inspector II

Examination II-1: Surveying Weaknesses

At this point in SAEP, you should have the process of self-directed learning using examinations fixed in your mind. Moving through Phase II is accomplished in the same way as in Phase I. Do not attempt to skip steps in the process because you understand how SAEP works. Skipping steps can lead to a weak examination-preparation result. The Phase II examinations will be more difficult because of the higher level of required knowledge and skills. You will find that the SAEP methods move you gradually from the simple to the complex.

Do not study prior to taking the examination. The examination is designed to identify your weakest areas in terms of NFPA Standard 1031. Some SAEP steps will eventually require self-study of specific reference materials.

Mark all answers in ink to ensure that no corrections or changes are made. Do not mark through answers or change answers in any way once you have selected the answer. Doing so indicates uncertainty regarding the answer. Mastery is not compatible with uncertainty.

Step 1—Take Examination II-1. When you have completed Examination II-1, compare your answers with the correct answers in Appendix B. Each answer cites relevant reference materials with page numbers. If you answered the examination item incorrectly, you have a source for conducting your correct answer research.

Step 2—Score Examination II-1. How many examination items did you miss? Write the number of missed examination items in the blank in ink ________________. Enter the number of examination items you guessed in this blank ________________. Enter these numbers in the designated locations in your Personal Progress Plotter.

Step 3—The learning begins! Carefully research the page cited in the reference material for the correct answer. For instance, use Delmar Thomson Learning *Fire Prevention, Inspection, and Code Enforcement*, second edition, go to the page number provided, and find the answer.

Following are some rules and helpful hints repeated from Phase I.

Rule 3

Mark with an "X" any examination items for which you guessed the answer. To obtain the maximum return on effort, research any answer that you guessed even if you guessed correctly. Find the correct answer, highlight it, and then read the entire paragraph that contains the answer. Be honest and mark all questions on which you guessed. Some examinations have a correction for guessing built into the scoring process. The correction for guessing can reduce your final examination score. If you are guessing, you are not mastering the material.

Rule 4

Read questions twice if you have any misunderstanding, especially if the question contains complex directions or activities.

Helpful Hint

Work through all missed examination items using the same technique. Reading the entire paragraph improves retention of the information and helps you develop an association with the material and learn the correct answers. This step may sound simple. A major finding during the development and field testing of SAEP was that you learn from your mistakes.

Examination II-1

Directions

Remove Examination II-1 from the manual. First, take a careful look at the examination. There should be 75 examination items. Notice that a blank line precedes each examination item number. Enter the answer to the examination item on the line. Write the answer in ink. Remember the rule about not changing your answers. Changed answers are often incorrect, and more often than not the answer that is chosen first is correct.

If you guess the answer to a question, place an "X" or a check mark by your answer. This step is vitally important as you gain and master knowledge. We will explain how we treat the "guessed" items later in SAEP.

Take the examination. Once you complete it, go to Appendix B and score your examination. Carefully follow the directions for feedback on the missed and guessed examination items.

_____ **1.** As witnesses in the courtroom, fire inspectors should confine their testimony to:
A. hearsay.
B. facts.
C. technical terms.
D. department policies.

_____ **2.** As advisors, fire inspectors can assist the prosecuting attorney with information about:
A. legal remedies.
B. criminal procedures.
C. immunity and equivalency.
D. fire ordinances and technical terms.

_____ **3.** When the local jurisdiction adopts an ordinance requiring an open burning permit, the inspector should give special attention to:
A. a list of individuals, property owners, and businesses that may apply for permits.
B. what, how, and where as well as conditions under which the permit is issued.
C. providing a fire watch for all sites to which permits for open burning are granted.
D. making clear his/her dissatisfaction with the parameters under which the permit was issued.

_____ **4.** An advantage of having an established plans review process is that:
A. it enables discrepancies to be fixed before construction begins.
B. inspectors can control the issuance of business licenses.
C. the inspector can take advantage of the contractor or the occupant.
D. construction costs can be estimated more accurately for the occupant.

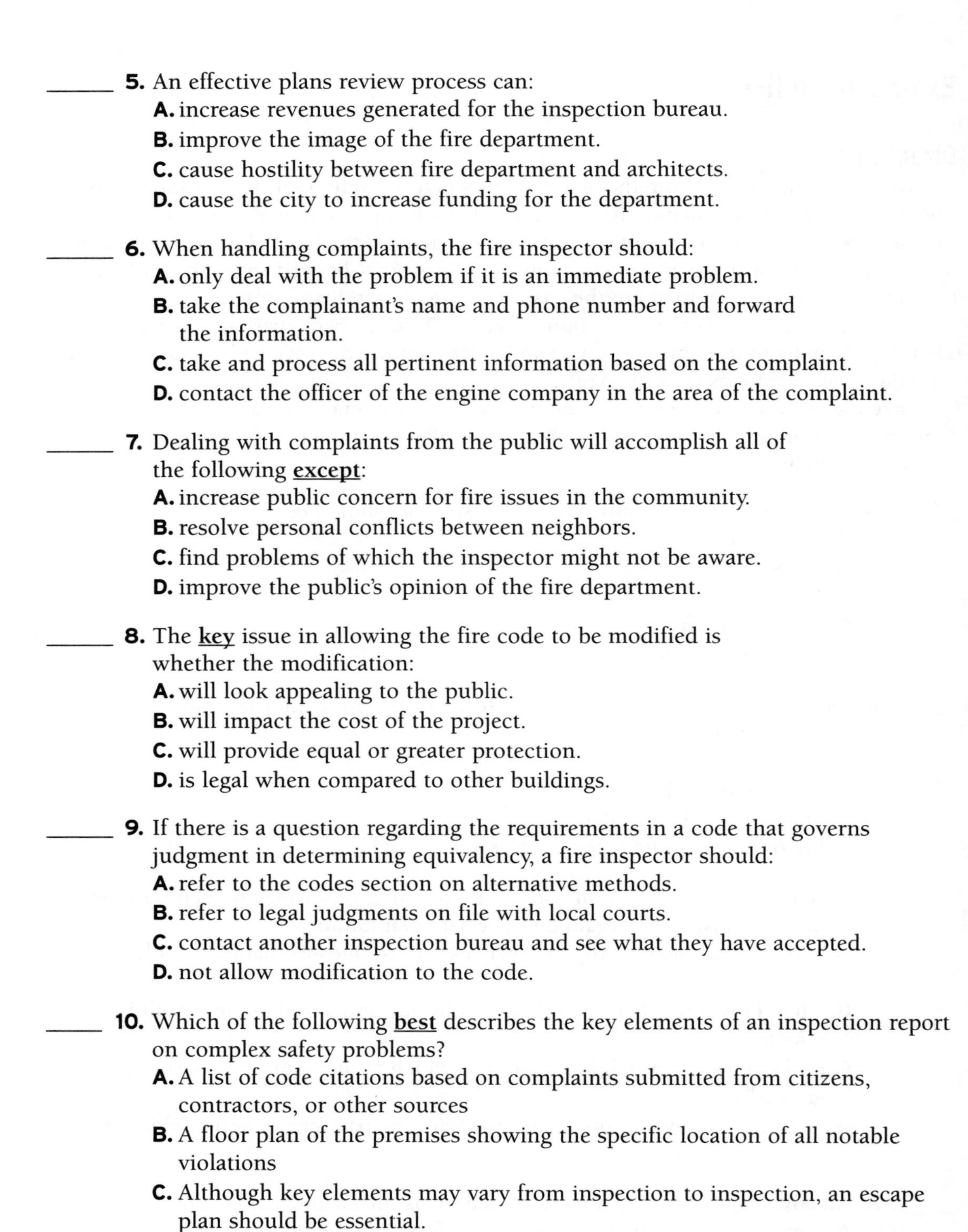

_____ **5.** An effective plans review process can:
A. increase revenues generated for the inspection bureau.
B. improve the image of the fire department.
C. cause hostility between fire department and architects.
D. cause the city to increase funding for the department.

_____ **6.** When handling complaints, the fire inspector should:
A. only deal with the problem if it is an immediate problem.
B. take the complainant's name and phone number and forward the information.
C. take and process all pertinent information based on the complaint.
D. contact the officer of the engine company in the area of the complaint.

_____ **7.** Dealing with complaints from the public will accomplish all of the following **except**:
A. increase public concern for fire issues in the community.
B. resolve personal conflicts between neighbors.
C. find problems of which the inspector might not be aware.
D. improve the public's opinion of the fire department.

_____ **8.** The **key** issue in allowing the fire code to be modified is whether the modification:
A. will look appealing to the public.
B. will impact the cost of the project.
C. will provide equal or greater protection.
D. is legal when compared to other buildings.

_____ **9.** If there is a question regarding the requirements in a code that governs judgment in determining equivalency, a fire inspector should:
A. refer to the codes section on alternative methods.
B. refer to legal judgments on file with local courts.
C. contact another inspection bureau and see what they have accepted.
D. not allow modification to the code.

_____ **10.** Which of the following **best** describes the key elements of an inspection report on complex safety problems?
A. A list of code citations based on complaints submitted from citizens, contractors, or other sources
B. A floor plan of the premises showing the specific location of all notable violations
C. Although key elements may vary from inspection to inspection, an escape plan should be essential.
D. Standard report format should be in written form and include statistical data on the business, pertinent dates, names, and phone numbers, list of violations and recommendations for correction of each violation, and date of follow-up inspection.

_____ **11.** A/an _______________ should be in written form and include statistical data on the business, dates, times, phone numbers, violations, recommendations for corrections, and follow-up inspection date.

A. inspection report
B. code requirement
C. enforcement procedure
D. permit

_____ **12.** Which one of the following statements <u>is not</u> true?

A. Written reports serve not only as records of an inspection, but may be needed in legal proceedings.
B. Written correction orders should always include the inspector's personal opinions.
C. Without written evidence of an inspection, no proof exists that the inspector gave the owner notice of hazardous conditions or corrective measures to be taken.
D. The inspection report left with the owner should inform, analyze, and recommend.

_____ **13.** Written reports serve not only as records of the inspection, but also can be used as a basis for:

A. a formal complaint.
B. legal action.
C. closing the file.
D. inspection drawings.

_____ **14.** Which of the following is the <u>correct</u> definition of occupant load?

A. The total number of persons that may safely occupy a building or portion thereof at any one time
B. The total number of persons that may exit a building
C. The total number of persons that may exit areas of concentrated use
D. The average number of persons that may exit a building or portion thereof at any one time
E. The maximum number of persons that may travel through or exit an occupied space at any one time

_____ **15.** When determining the occupant load for a multi-use building, the fire inspector should determine the occupant load based on:

A. the occupancy that has the greatest risk factor.
B. the occupancy that has the least number of persons.
C. the occupancy that has the greatest number of persons.
D. the total area of the building regardless of occupancy.

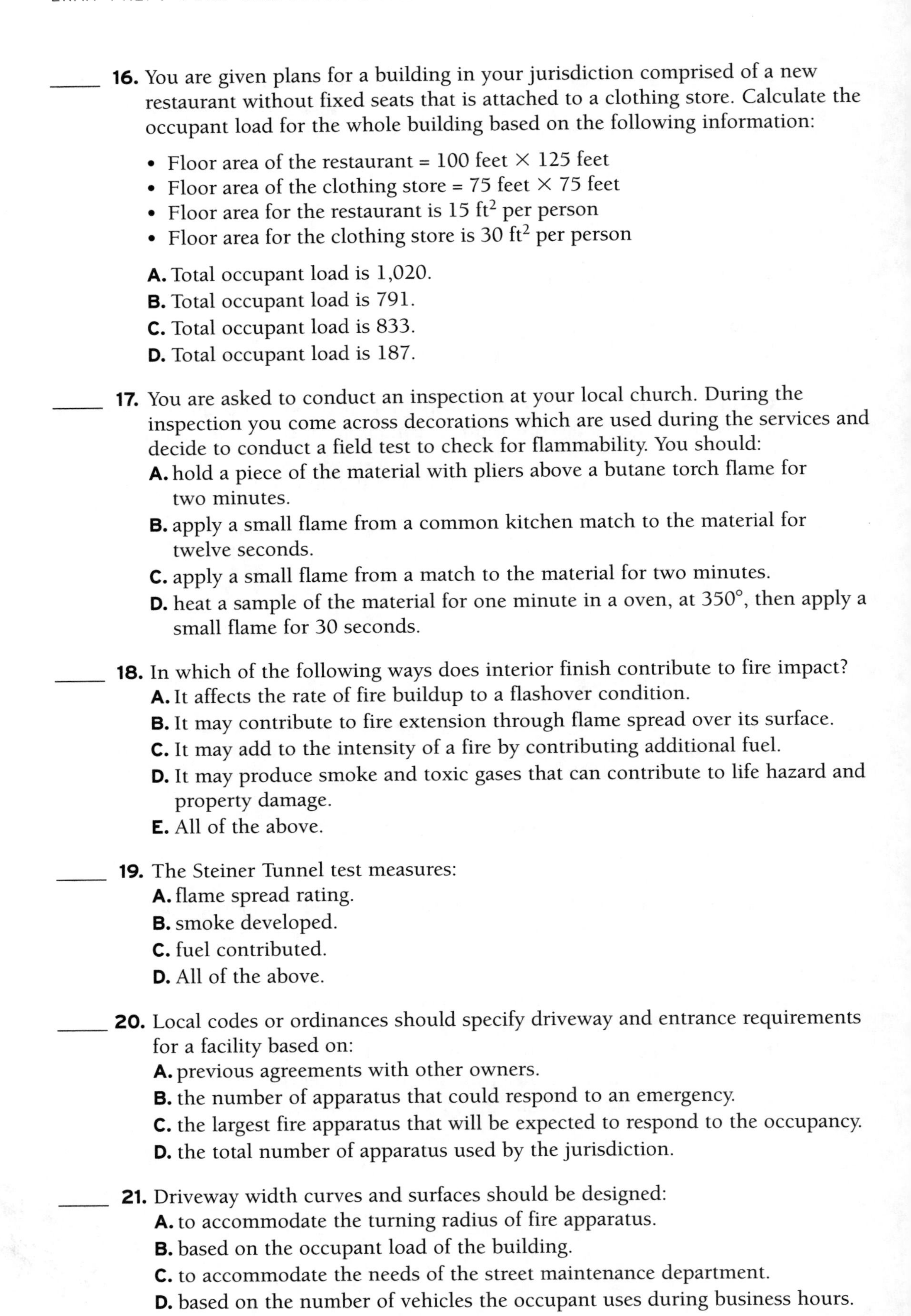

_____ **16.** You are given plans for a building in your jurisdiction comprised of a new restaurant without fixed seats that is attached to a clothing store. Calculate the occupant load for the whole building based on the following information:

- Floor area of the restaurant = 100 feet × 125 feet
- Floor area of the clothing store = 75 feet × 75 feet
- Floor area for the restaurant is 15 ft^2 per person
- Floor area for the clothing store is 30 ft^2 per person

A. Total occupant load is 1,020.
B. Total occupant load is 791.
C. Total occupant load is 833.
D. Total occupant load is 187.

_____ **17.** You are asked to conduct an inspection at your local church. During the inspection you come across decorations which are used during the services and decide to conduct a field test to check for flammability. You should:
A. hold a piece of the material with pliers above a butane torch flame for two minutes.
B. apply a small flame from a common kitchen match to the material for twelve seconds.
C. apply a small flame from a match to the material for two minutes.
D. heat a sample of the material for one minute in a oven, at 350°, then apply a small flame for 30 seconds.

_____ **18.** In which of the following ways does interior finish contribute to fire impact?
A. It affects the rate of fire buildup to a flashover condition.
B. It may contribute to fire extension through flame spread over its surface.
C. It may add to the intensity of a fire by contributing additional fuel.
D. It may produce smoke and toxic gases that can contribute to life hazard and property damage.
E. All of the above.

_____ **19.** The Steiner Tunnel test measures:
A. flame spread rating.
B. smoke developed.
C. fuel contributed.
D. All of the above.

_____ **20.** Local codes or ordinances should specify driveway and entrance requirements for a facility based on:
A. previous agreements with other owners.
B. the number of apparatus that could respond to an emergency.
C. the largest fire apparatus that will be expected to respond to the occupancy.
D. the total number of apparatus used by the jurisdiction.

_____ **21.** Driveway width curves and surfaces should be designed:
A. to accommodate the turning radius of fire apparatus.
B. based on the occupant load of the building.
C. to accommodate the needs of the street maintenance department.
D. based on the number of vehicles the occupant uses during business hours.

_____ **22.** The <u>first</u> design element of a fire department access road surface should be:
A. clear of all debris at all times.
B. clear of snow or hazardous conditions.
C. wide enough to handle two apparatus side by side.
D. strong enough to handle the weight of a fire apparatus.

_____ **23.** A <u>primary</u> fire hazard associated with most equipment fires is:
A. combustibles are too close.
B. improper grounding.
C. hazardous chemicals.
D. fuel.

_____ **24.** In high rise buildings, enclosed stairwells are designed with _______________ and _______________ requirements.
A. pressurization and occupancy
B. fire rating and ADA (Americans with Disabilities Act)
C. fire rating and pressurization
D. egress and ADA (Americans with Disabilities Act)

_____ **25.** Construction classifications are based on:
A. regional locations.
B. authority having jurisdiction.
C. the materials that are contained in the building.
D. the materials used in construction of the building.

_____ **26.** Smoke-proof enclosures provide the highest degree of fire protection for stair enclosures. In buildings that are four stories or higher, they must be enclosed from highest point to lowest point by fire barriers. The rating must be **<u>at least</u>**:
A. 1 hour.
B. 2 hours.
C. 3 hours.
D. 4 hours.

_____ **27.** In new building construction, a door opening serving as a means of egress must be at least _______________ clear width.
A. 32"
B. 30"
C. 48"
D. 36"

_____ **28.** Fire inspectors can estimate the fire load when conducting inspections of a building. This information may be used for all of the following **<u>except</u>**:
A. fire extinguisher placement.
B. prefire plans.
C. occupant load.
D. water requirements for automatic sprinklers.

_____ **29.** Draft curtains are sometimes used in large areas of buildings that **are not** otherwise subdivided. The purpose of a draft curtains is to:
A. limit the mushrooming effect of heat and smoke.
B. help with the air-conditioning needs of the area.
C. deflect hose streams into hard to reach areas.
D. stop the spread of fire through a fire assembly.

_____ **30.** Carbon dioxide fixed extinguishing systems may be used in areas where fires could occur in:
A. flammable liquids and gases.
B. rubber tire storage.
C. paper storage.
D. combustible metals.

_____ **31.** Dry chemical fixed fire extinguishing systems **are not** recommended for use on:
A. flammable liquids.
B. flammable gases.
C. delicate electronic equipment.
D. paper storage.

_____ **32.** Which of the following are types of fixed fire extinguishing systems?
A. Automatic sprinkler systems, foam systems, carbon dioxide systems, halogenated agent systems, and chemical systems
B. Automatic sprinkler systems, factory mutual systems, and U.L. listed systems
C. Wet sprinkler systems, dry sprinkler systems, deluge systems, fire mutual systems, and hazardous substance systems
D. Water sprinkler systems, air sprinkler systems, flooding sprinkler systems, and chemical sprinkler systems

_____ **33.** To ensure that sprinkler systems perform properly during a fire, they require:
A. maintenance, periodic inspection, and F.M. Listing.
B. periodic inspection, testing, and U.L. Listing.
C. U.L. Listing, testing, and maintenance.
D. periodic inspection, testing, and maintenance.

_____ **34.** Which of the following are parts of a fire door assembly?
A. Rated fire door
B. Rated frame assembly
C. Latching hardware
D. Door closing hardware
E. All of the above

_____ **35.** Standard fire doors include which of the following types?
A. Horizontal sliding
B. Vertical sliding
C. Overhead rolling
D. Double swinging
E. All of the above

_____ **36.** Prior to pumping, the amount of pressure that is expected to be available from a hydrant is called normal _________________ pressure.
A. atmospheric
B. operating
C. residual
D. grid

_____ **37.** Which of the following **is not** a primary water supply source for an automatic sprinkler system?
A. Public water-supply system
B. Gravity tank
C. Pressure tank
D. Fire department water tanker truck

_____ **38.** Two causes of increased resistance or friction loss in water mains are:
A. encrustations and conduction.
B. sedimentation and encrustation.
C. elasticity and conduction.
D. sedimentation and elasticity.

_____ **39.** _________________ will completely line the interior of a pipe and will gradually cause a restricted diameter.
A. Lead deposits
B. Sedimentation
C. Encrustations
D. Mud deposits

_____ **40.** _________________ deposits consist of mud, clay, and dead organisms.
A. Valve
B. Sedimentary
C. Tuberculation
D. Encrustation

_____ **41.** A fire hydrant that receives water from only one direction is called a:
A. one-way hydrant.
B. steamer hydrant.
C. circulating-feed hydrant.
D. dead-end hydrant.

_____ **42.** It is very important that, during plans review of a building, the Fire Inspector is knowledgeable in _________________ and _________________.
A. sprinkler systems, fire alarms
B. zoning, sprinkler systems
C. fire alarms, zoning
D. planning, fire alarms

_____ **43.** Which of the following are part of a means of egress?
A. Corridor, stairs, alley
B. Overhead door, non-rated corridor, alley
C. Through kitchen, corridor, street
D. Through mechanical room, fenced in yard, street

_____ **44.** A means of egress consists of three distinct components:
A. exit access, exit, and exit discharge.
B. exit, exit access, and enclosed hallway.
C. exit discharge, passage way, and ramp.
D. escalator, fire escape, and stairs.

_____ **45.** When panic hardware is required, occupants should be able to open the door by applying a force of **not more than** _______________ pounds.
A. 5
B. 10
C. 15
D. 20

_____ **46.** When illumination of exits is required, it must be continuous during occupancy and floors must be illuminated at not less than 1 footcandle (10.8 Lumens (1X) measured at the:
A. floor.
B. ceiling.
C. wall.
D. ceiling and floor.

_____ **47.** When two or more exits from an occupancy are required, they shall be located:
A. remote from each other.
B. remote as long as travel distance is not too long.
C. remote as to provide same travel distance for 80% of the occupants.
D. an equal distance from all occupants.

_____ **48.** Cooking equipment must have a clearance of at least _______________ inches to any combustible material.
A. 6
B. 10
C. 12
D. 18

_____ **49.** It is essential, in order for fire inspectors to be effective, that they know:
A. the hazards involving equipment and processes found in their jurisdiction.
B. all codes involving all types of equipment or processes.
C. only the codes based on past problems with which the jurisdiction has dealt.
D. the hazards involved in industrial occupancies.

_____ **50.** A key element of an emergency plan, common to all types of occupancies is:
A. an approved inspector's report.
B. the need to evacuate the structure in a timely and efficient manner.
C. notifying the fire department of any emergency.
D. making sure all areas of the building are accessible.

_____ **51.** School fire drills should be conducted:
A. at different times of the day.
B. only when it is convenient for the teachers.
C. at the beginning of classes.
D. at the discretion of the local fire department.

_____ **52.** Tanks designed to hold large quantities of low pressure gas are built according to standards set forth by the:
A. U.S. Department of Transportation.
B. Boiler and Pressure Vessel Code.
C. American Petroleum Institute (API).
D. American National Standards Institute (ANSI).

_____ **53.** Which of the following is a poor handling or storage practice involving gases?
A. Storage of cylinders in non-ventilated areas
B. Securing cylinders properly or using a cap
C. Providing a pressure relief device with appropriate settings and adequate pressure relief capacity
D. Storing of compatible gases in the same area

_____ **54.** When referring to the L.E.L. of a flammable or combustible liquid, what does the L.E.L. mean?
A. Lowest Efficiency Level
B. Lower Exposure Label
C. Lower Explosive Limit
D. Lowest Evaporation Limit

_____ **55.** Portable tanks with capacity greater than _______________ gallons should be treated according to the same standards as fixed tanks.
A. 60
B. 120
C. 240
D. 660

_____ **56.** The storage magazine required for black powder and smokeless powder is a Type _______________ Magazine.
A. 1
B. 2
C. 3
D. 4
E. 5

_____ **57.** The <u>best</u> source of information on a specific hazardous material from the manufacturer is the:
A. Material Safety Data Sheet (MSDS).
B. Coast Guard Hazard Response Information System Manual (CHRIS Manual).
C. Bureau of Alcohol, Tobacco, and Firearms (ATF).
D. American Association of Rail Roads (AARR).

_____ **58.** Buildings or structures used primarily for securing and sheltering goods, merchandise, products, vehicles, or animals are classified as _______________ occupancies.
A. storage
B. mercantile
C. business
D. warehouse

_____ **59.** Buildings are classified by _______________ according to the type of occupancy or intended end use.
A. NFPA 1710
B. NFPA 101, Life Safety Code
C. NFPA 10, Fire Prevention Code
D. ISO Insurance Services Office Code

_____ **60.** When exits serve more than one floor, an important element in determining the exit capacity is the:
A. maximum square feet per floor.
B. overall height allowable.
C. maximum occupant load per floor.
D. All of the above.

_____ **61.** If a building is 80 feet long and 120 feet wide and is used as a dance hall (7 sq. ft. per person), the occupancy load for the building is _______________ persons.
A. 1,370
B. 1,730
C. 648
D. 684

_____ **62.** Occupant load for an assembly occupancy without fixed seating (15 $ft.^2$ per person) that is 100 feet × 200 feet is _______________ persons.
A. 1,000
B. 1,233
C. 1,333
D. 1,323

_____ **63.** Fire protection engineers have established _______________ psi as the minimum required residual pressure for a water supply system.
A. 10
B. 15
C. 20
D. 25

_____ **64.** When the water source <u>**does**</u> <u>**not**</u> have adequate elevation to create proper pressure for gravity flow, it is necessary to use:
A. larger size pipes in the mains.
B. pumps to raise the system's pressure.
C. computer controlled pressure regulators.
D. negative coefficient of friction loss.

_____ **65.** When a fire hydrant receives water from two or more directions, it is said to have _______________ feed.
A. compound
B. circulating
C. compensating
D. distributed

_____ **66.** Currently, _______________-inch pipe is becoming a **minimum** size used for a water distribution system.
A. 4
B. 6
C. 8
D. 12

_____ **67.** NFPA 101, Life Safety Code requires that a Class A ramp must be at least _______________ inches wide.
A. 22
B. 36
C. 30
D. 44

_____ **68.** When an opening in a three-hour fire-rated wall needs a fire-rated door, the requirement should be for:
A. 2 two-hour doors on one side.
B. three-hour doors, one on each side.
C. 2 two-hour doors, one on each side.
D. 1 three-hour door.

_____ **69.** Building construction techniques and materials are tested under two NFPA standards. What are they?
A. NFPA 255 and 257
B. NFPA 251 and 255
C. NFPA 252 and 257
D. NFPA 257 and 251

_____ **70.** A building's structural members including walls, columns, beams, floors, and roofs are of noncombustible or limited combustible materials. The degree of fire resistance of this building is:
A. Type IV.
B. Type I.
C. Type II.
D. Type III.

_____ **71.** According to NFPA 220, Standard on Types of Building Construction, Type II construction is similar to Type I construction **except** that the degree of fire resistance is:
A. lower.
B. higher.
C. fire rated.
D. approved by NFPA.

_____ **72.** In regard to fire walls, the type of construction, the size of the area protected, and the severity of the fire hazard will govern the degree of fire resistance required, which generally ranges from _______________ hours.
A. 1 to 4
B. 3/4 to 2
C. 2 to 4
D. 3/4 to 4

_____ **73.** According to NFPA 220, Standard on Types of Building Construction, Type I construction has structural members including walls, floors, columns, beams, and roofs of ________________ or ________________ materials.

A. steel, laminated wood

B. noncombustible, limited combustible

C. wood, heavy timber

D. heavy timber, steel

_____ **74.** The flammability of roof materials are classified as:

A. Class I, Class II, and Class III.

B. Type I, Type II, and Type III.

C. Class A, Class B, and Class C.

D. Type A, Type B, and Type C.

_____ **75.** In general, construction classifications are based upon materials used in construction and upon hourly fire-resistance ratings of:

A. the National Fire Code.

B. interior finish.

C. structural components.

D. interior design.

Did you score higher than 80 percent on Examination II-1? Circle Yes or No in ink. (We will return to your answer to this question later in SAEP.)

Examination II-2: Adding Difficulty and Depth

During Examination II-2, progress will be made in developing depth of knowledge and skills. Follow the steps carefully to realize the best return on effort.

Step 1—Take Examination II-2. When you have completed Examination II-2, compare your answers with the correct answers in Appendix B.

Step 2—Score Examination II-2. How many examination items did you miss? Write the number of missed examination items in the blank in ink ________________. Enter the number of examination items you guessed in this blank ________________. Enter these numbers in the designated locations in your Personal Progress Plotter.

Step 3—The learning begins. During the feedback step, use Appendix B information for Examination II-2 to research the correct answer for items you missed or guessed. Highlight the correct answer during your research of the reference materials. Read the entire paragraph that contains the correct answer.

Examination II-2

Directions

Remove Examination II-2 from the manual. First, take a careful look at the examination. There should be 75 examination items. Notice that a blank line precedes each examination item number. Enter the answer to the examination item on this line. Write the answer in ink. Remember the rule about not changing your answers. Changed answers are often incorrect, and more often than not the answer that is chosen first is correct.

If you guess the answer to a question, place an "X" or a check mark by your answer. This step is vitally important as you gain and master knowledge. We will explain how we treat the "guessed" items later in SAEP.

Take the examination. Once you complete it, go to Appendix B and score your examination. Carefully follow the directions for feedback on the missed and guessed examination items.

_____ **1.** The nature of code enforcement is complex, and inspectors often need help from:
A. outside technical assistance.
B. judges.
C. firefighters.
D. the fire chief.

_____ **2.** In court, fire inspectors should remain impartial and calm, never entering into _______________ with the attorney for the defendant.
A. dissertations
B. technical terminology
C. arguments
D. semantics

_____ **3.** The **first** step in the permit process is the:
A. fire inspector conducting an inspection for the permit.
B. occupant contacting the fire inspector and determining whether a permit is needed.
C. inspection department receiving a complaint.
D. fire inspector issuing a permit over the phone.

_____ **4.** Part of the permit process includes completing an application and providing documentation. This documentation includes all of the following **except**:
A. shop drawings.
B. plot diagrams.
C. MSDS or other chemical documentation.
D. a preplan.

_____ **5.** If an agency **does not** have a fire protection engineer employed to conduct plan reviews, the responsibility:
A. is left to the building or zoning official.
B. does not have to be completed.
C. is assigned to a fire inspector.
D. is given to the company officer that does company inspections for the area.

_____ **6.** If given the assignment to conduct a plans review, fire inspectors should:
A. have a complete knowledge of construction techniques.
B. know applicable codes for their jurisdiction.
C. know their legal limitations.
D. All of the above.

_____ **7.** When investigating a complaint, a fire inspector should anticipate that the occupant/owner may:
A. be cooperative and helpful.
B. be resistive.
C. deny the inspector access.
D. All of the above.

_____ **8.** When investigating a complaint, the fire inspector:
A. is not required to give the owner advance notice.
B. should call the owner and set an appointment.
C. should be accompanied by the complainant.
D. should be accompanied by the police.

_____ **9.** The power given to the inspector to modify a fire code is determined by the:
A. National Fire Code.
B. International Conference of Building Codes.
C. National Fire Protection Association.
D. authority having jurisdiction.

_____ **10.** Which is not a concern when an inspector is completing an inspection report?
A. Proving the evidence
B. Presenting the facts
C. Recording the inspector's personal position
D. Justification of a recommendation

_____ **11.** In most cases, a formal inspection report will only be necessary:
A. when those inspections reported are on a standard checklist.
B. when site drawings have not been provided.
C. in cases of life-threatening hazards, major renovations, and/or an extensive list of violations.
D. when the inspector is unsure of a special requirement.

_____ **12.** Which one of the following resources would not be a good source for creating a checklist for the inspection of various occupancies?
A. NFPA - Fire Protection Handbook
B. NFPA - Conducting Fire Inspections, A Guide for Field Use
C. NFPA - Life Safety Code Handbook
D. NFPA - Field Incident Guide

_____ **13.** To determine the occupant load for a building that contains two or more distinct occupancies, the total occupant load for the building is determined by calculating the occupant load:
A. based on the occupancy that is the largest in square footage.
B. based on the occupancy that has the highest degree of hazard.
C. based on the occupancy that has the greatest means of egress.
D. of each occupancy separately and then adding them together.

_____ **14.** Using the floor area method, what is the occupant load factor for a structure 100 × 150 feet, one story, with two 36-inch wide exits for existing business occupancy?
A. 100
B. 150
C. 50
D. 200

_____ **15.** The single unit of exit width used to determine occupant load is _______________ inches.
A. 12
B. 18
C. 22
D. 24
E. 36

_____ **16.** A chemical designed to retard ignition or the spread of fire when it is applied to material or another substance is defined as a:
A. fire proofer.
B. flame arrester.
C. fire rating.
D. fire retardant.

_____ **17.** A test designed to determine the surface burning characteristics of interior finishes is called a _______________ test.
A. carpet-pill
B. fire spread
C. Steiner tunnel
D. flame spread

_____ **18.** A numerical rating assigned to a material based on the speed and extent to which flame travels over its surface is called a _______________ rating.
A. burning ember
B. fire resistance
C. flashover
D. flame spread

_____ **19.** During a field inspection, the fire inspector should do all of the following <u>**except**</u>:
A. ensure that overhead obstruction will not interfere with fire apparatus placement.
B. ensure that the road surface of a facility will handle the weight of an apparatus.
C. determine the cost to the contractor of changes being required by the fire department.
D. ensure that the entrance to the facility is the correct width.

_____ **20.** During a routine inspection, it is determined that the road surface to a facility is inadequate to handle the weight of an apparatus. The inspector should:
A. note the problem on the inspection form.
B. notify suppression personnel that would respond to the location.
C. close the facility down until it is corrected.
D. Both A and B.
E. Both A and C.

_____ **21.** During plans review, an inspector should determine _______________ for apparatus access.
A. surface runoff
B. turning radius
C. curb height
D. parking space

_____ **22.** Smoke control systems in new construction are designed to be used in conjunction with:
A. fire and smoke dampers.
B. fire doors and partitions.
C. curtain boards (draft curtains).
D. All of the above.

_____ **23.** In performance-based design, the acceptance test criteria, methods of evaluating data, and approval procedures are developed by the:
A. architect and contractor.
B. building owner.
C. authority having jurisdiction.
D. planning commission.
E. All of the above.

_____ **24.** How a building is designed is referred to as its _______________ classification.
A. occupancy
B. zoning
C. construction
D. fire code

_____ **25.** What type of exit enclosure provides the highest degree of protection of stair enclosures?
A. A rated exit door enclosure
B. A continuous exit enclosure
C. A smoke proof enclosure
D. A horizontal exit enclosure

_____ **26.** A smoke-proof enclosure in a three story building must be enclosed from its highest point to its lowest point by a _________________-hour fire rating.
A. one
B. two
C. three
D. four

_____ **27.** The upright-type of sprinkler head:
A. may only be used in a wet-pipe configuration.
B. protrudes downward from the exposed pipe.
C. cannot be inverted for use in the pendant-type sprinkler head.
D. can be installed in a side wall application.

_____ **28.** The NFPA standard which requires that fire extinguishers be thoroughly inspected is:
A. NFPA 25.
B. NFPA 14.
C. NFPA 13.
D. NFPA 10.

_____ **29.** Sprinkler systems shall be inspected, tested, and maintained in accordance with:
A. NFPA 10.
B. NFPA 13.
C. NFPA 25.
D. NFPA 14.

_____ **30.** A **minimum** clearance of _________________ inches shall be maintained below each ordinary hazard/standard spray sprinkler head as measured from the deflector.
A. 6
B. 12
C. 18
D. 24

_____ **31.** All maintenance procedures should include a thorough examination of the three basic parts of a portable extinguisher:
A. mechanical parts, extinguishing agent, and expelling means.
B. mechanical parts, color, and expelling means.
C. gauge, size, and color.
D. extinguishing agent, nozzle, and gauge.

_____ **32.** A network of intermediate-sized pipe that reinforces the overall grid system by forming loops that interlock primary feeders **best** defines:
A. primary loop.
B. secondary feeders.
C. distributors.
D. grid network.

_____ **33.** Large pipes which carry large quantities of water to various points along the water supply system for distribution to smaller mains **best** defines:
A. primary feeders.
B. secondary feeders.
C. distributors.
D. grid network.

_____ **34.** Today, 8" pipe is becoming the minimum size pipe in a water distribution system because:
A. of its higher flow capacity advantages over smaller diameter pipe.
B. all pipe would then be the same diameter.
C. the pipe size determines the flow pressures needed to design a reliable system.
D. new code requirements are phasing out small diameter pipe.

_____ **35.** The _________________ system is equipped with a dry type valve and all sprinkler heads are open.
A. wet-pipe
B. dry-pipe
C. deluge
D. pre-action

_____ **36.** The hose connections to the FDC (Fire Department Connection) may be of the _________________ type and equipped with standard caps.
A. right-handed
B. left-handed
C. female
D. male

_____ **37.** The standard hydrostatic test for all piping in a wet pipe systems is not less than _________________ for _________________.
A. 50 psi, 1 hour
B. 200 psi, 2 hours
C. 150 psi, 20 minutes
D. 100 psi, 1-1/2 hour

_____ **38.** Which of the following **would not** be considered an obsolete portable fire extinguisher?
A. Soda-acid
B. Water cartridge operated
C. Vaporization liquid
D. Stored pressure

_____ **39.** The color of a sprinkler head frame that has a temperature classification of "High" is:
A. red.
B. blue.
C. green.
D. orange.

_____ **40.** According to NFPA 25, the heat for the dry-valve enclosure should be checked _________________ during freezing weather.
A. daily
B. weekly
C. monthly
D. quarterly

_____ **41.** Which one of the following **would** **not** cause reduction of a sprinkler system's effectiveness?
A. Opening of nearby hydrants
B. Freezing of the gravity tank
C. Storage piled around heads
D. Pumping into the system

_____ **42.** The rating required for a fire door in an exit stairway is determined by:
A. where the exit empties.
B. whether it is a ramp or stairway.
C. the fire rating of the enclosure.
D. All of the above.

_____ **43.** In addition to determining how many people may occupy a building, an inspector must determine the number of exits. This is part of the:
A. load capacity.
B. means of egress.
C. means of access.
D. means of exit.

_____ **44.** The components of a means of egress include:
A. exit access.
B. the exit.
C. exit discharge.
D. All of the above.

_____ **45.** According to NFPA 101, Life Safety Code, exit stairs are a critical component of the means of egress in a multi-storied building. Stairways must be at least _________________ inches wide, unless the total occupant load of all the floors served by the stairways is less than 50. In this case, stairways may be _________________ inches wide.
A. 44, 24
B. 48, 36
C. 36, 24
D. 36, 32
E. 44, 36

_____ **46.** The four main views of working drawings are:
A. floor plan, mechanical, electrical, and fire protection.
B. fire protection, elevation, sectional, and floor plan.
C. plan, elevation, sectional, and detailed.
D. mechanical, sectional, elevation, and site plan.

_____ **47.** There are inherent hazards associated with dip tanks; therefore, dip tank operations:
A. have a smoke detector tied into a fire alarm system.
B. should never be placed in a basement.
C. must not have overflow devices on the tank.
D. must never be covered when not in use.
E. must be located below ground.

_____ **48.** Static charges may be generated by all of the following **except**:
A. belts in motion.
B. magnetic fields.
C. moving vehicles.
D. non-conductive fluids flowing through pipes.

_____ **49.** The vents used for underground storage tanks should have at least a _______________ inch nominal inside diameter.
A. 2
B. 1-1/2
C. 1-1/4
D. 1

_____ **50.** Hotels and motels must provide different evacuation procedures because of:
A. the number of floors.
B. the shape of the building.
C. the multiple uses of the building.
D. temporary occupancy.

_____ **51.** The facilities that do not warrant total evacuation every time the fire alarm sounds are:
A. daycare centers, schools, and hotels.
B. hospitals and nursing homes.
C. hospitals, schools, and daycare centers.
D. hotels, motels, and nursing homes.

_____ **52.** Which is a hazard associated with the handling and use of flammable and combustible liquids?
A. Ventilation
B. Static electricity
C. Closed containers
D. Bonding containers

_____ **53.** Flammable or combustible liquid containers stored in warehouses or buildings must be arranged to provide access aisles of approximately _______________ in width.
A. 3 feet
B. 6 feet
C. 9 feet
D. 12 feet

_____ **54.** The maximum allowable size for approved safety cans for flammable or combustible liquids is _______________ gallon(s).
A. 2.5
B. 2
C. 1
D. 5

_____ **55.** Fuel oil storage tanks inside of buildings require no special fire protection features if tank capacity is less than _______________ gallons.
A. 660
B. 1100
C. 30
D. 500

_____ **56.** On above-ground inside storage tanks for flammable or combustible liquid storage tanks, automatic closing, heat-actuated valves are required on:
A. all piping connections.
B. piping connections above the liquid level.
C. none of the piping connections.
D. all piping connections below the liquid level.

_____ **57.** Above ground atmospheric tanks for flammable or combustible liquids are designed for pressures of _______________ psi.
A. 0 to 0.5
B. 0.5 to 15
C. 15 to 20
D. 10 to 15

_____ **58.** The best source of information on a specific hazardous material from the manufacturer is the:
A. Material Safety Data Sheet (MSDS).
B. Coast Guard Hazard Response Information System Manual (CHRIS Manual).
C. Bureau of Alcohol, Tobacco, and Firearms (ATF).
D. American Association of Rail Roads (AARR).

_____ **59.** An identification system that can be implemented in an area where hazardous materials are regularly stored and used is the:
A. United Nations numbering system.
B. North American Emergency Response Guide.
C. NFPA 704 system.
D. Material Safety Data Sheets.

_____ **60.** A document that allows the inspector to match a product with the recommended hazard level is:
A. NIOSH 1910.120.
B. 49 CFR 171.8.
C. D.O.T Response Guide.
D. NFPA 49.

_____ **61.** Stores, markets, and other rooms, buildings, or structures used to display and sell merchandise are classified as _______________ occupancies.
A. business
B. commercial
C. retail
D. mercantile

_____ **62.** The classification given to a particular building based on the materials and methods used to construct and the ability of those materials to resist the effects of a fire situation is the _______________ classification.
A. construction
B. occupancy
C. flame spread
D. fire hazard

_____ **63.** The obvious advantage of establishing a plans review process is that it:
A. allows the inspector to spend more time in the office.
B. creates an additional opportunity for training and fire inspection.
C. enables the reviewer to point out discrepancies before construction begins.
D. increases the need for fire suppression activities.

_____ **64.** A floor plan sketch or diagram of a structure consists of:
A. a detailed view of the construction features of each floor of a building.
B. an outline of each floor of the building showing walls, partitions, and openings.
C. a sectional view of a particular portion of a building.
D. showing the building and grounds as they are actually seen in depth by the eye.

_____ **65.** A dance floor located in a drinking establishment measures 18 feet by 24 feet. Calculate the occupant load using the floor area method using 7 square feet per person.
A. 9
B. 15
C. 29
D. 62
E. 69

_____ **66.** The smallest of the mains in a water distribution system that serve the individual fire hydrants and blocks of consumers are called:
A. indicating feeders.
B. secondary feeders.
C. primary feeders.
D. distributors.

_____ **67.** Valve spacing for water systems in high value districts should be no more than _______________ feet.
A. 300
B. 500
C. 750
D. 1,000

_____ **68.** Which of the following would not normally be part of a grid system?
A. Primary feeders
B. Secondary feeders
C. Inter-connecting distributors
D. Dead-end water mains

_____ **69.** Which fire resistance rating is not a standard time rating?
A. 5 hours
B. 1 hour
C. 45 minutes
D. 4 hours

_____ **70.** The roof covering that provides the best fire retardant properties is:
A. Class A.
B. Class B.
C. Class C.
D. Class D.

_____ **71.** The fire resistance of a structural component is a function of various properties of the materials used, including:
A. combustibility.
B. thermal conductivity.
C. chemical composition.
D. dimensions.
E. All of the above.

_____ **72.** An office building whose structural members including walls, columns, beams, floors, and roof are of noncombustible or limited combustible materials and whose fire resistance rating of the structural members is of the highest fire resistance rating is _______________ construction.
A. Type I
B. Type II
C. Type III
D. Type IV
E. Type V

_____ **73.** According to NFPA 256, Methods of Fire Tests of Roof Coverings, the roof covering that provides the least fire retardant properties is:
A. Class A.
B. Class B.
C. Class C.
D. Class D.

_____ **74.** Heavy timber construction is allowed for interior structural members in _______________ construction.

A. Type I
B. Type II
C. Type III
D. Type IV
E. Type V

_____ **75.** Which of the following tests **is not** used to classify the flammability of roof coverings?

A. Intermittent flame exposure test
B. Burning brand test
C. Steiner tunnel test
D. Flame spread test

Did you score higher than 80 percent on Examination II-2? Circle Yes or No in ink. (We will return to your answer to this question later in SAEP.)

Examination II-3: Surveying Weaknesses and Improving Examination-Taking Skills

This examination section is designed to identify your remaining weaknesses in areas covered by NFPA Standard 1031, Fire Inspector II. Examination II-3 is randomly generated and contains both examination items you have taken before and new items. Some steps in SAEP will require self-study of specific reference materials.

Mark all answers in ink to ensure that no corrections or changes are made. Do not mark through answers or change answers in any way once you have selected your answer.

Step 1—Take Examination II-3. When you have completed Examination II-3, compare your answers with the correct answers in Appendix B. Notice that reference materials with page numbers are provided for each answer. If you missed the examination item, you have a source for researching the correct answer.

Step 2—Score Examination II-3. How many examination items did you miss? Write the number of missed examination items in the blank in ink ________________. Enter the number of examination items you guessed in this blank ________________. Enter these numbers in the designated locations on your Personal Progress Plotter.

Step 3—Reinforce what you have learned! During the feedback step, research the correct answer using the Appendix B information for Examination II-3. Highlight the correct answer during your research of the reference materials. Read the entire paragraph containing the correct answer.

Examination II-3

Directions

Remove Examination II-3 from the manual. First, take a careful look at the examination. There are 100 examination items. Notice that a blank line precedes each examination item number. Enter your answer to the examination item on this line. Write the answer in ink. Remember the rule about not changing your answers. Changed answers are most often incorrect, and more often than not the one that is chosen first is correct.

If you guess the answer to a question, place an "X" or check mark by your answer. This step is vitally important for gaining and mastering knowledge. We will explain how we treat the "guessed" items later in SAEP.

Take the examination. Once you complete it, go to Appendix B and score your examination. Carefully follow the directions for feedback on the missed and guessed examination items.

_____ **1.** As witnesses in the courtroom, fire inspectors should confine their testimony to:
A. hearsay.
B. facts.
C. technical terms.
D. department policies.

_____ **2.** As advisors, fire inspectors can assist the prosecuting attorney with information about:
A. legal remedies.
B. criminal procedures.
C. immunity and equivalency.
D. fire ordinances and technical terms.

_____ **3.** In court, fire inspectors should remain impartial and calm, never entering into _______________ with the attorney for the defendant.
A. dissertations
B. technical terminology
C. arguments
D. semantics

_____ **4.** A legal reprimand/charge for failure to comply with laws or regulations is a/an:
A. statute.
B. citation.
C. complaint.
D. ordinance.

_____ **5.** In order to require open-burning permits, a local municipality should:
A. enact a policy.
B. adopt an ordinance.
C. write a resolution.
D. write a procedure.

_____ **6.** The <u>first</u> step in the permit process is the:
A. fire inspector conducting an inspection for the permit.
B. occupant contacting the fire inspector and determining whether a permit is needed.
C. inspection department receiving a complaint.
D. fire inspector issuing a permit over the phone.

_____ **7.** An advantage of having an established plans review process is that:
A. it enables discrepancies to be fixed before construction begins.
B. inspectors can control the issuance of business licenses.
C. the inspector can take advantage of the contractor or the occupant.
D. construction costs can be estimated more accurately for the occupant.

_____ **8.** If an agency <u>does</u> <u>not</u> have a fire protection engineer employed to conduct plan reviews, the responsibility:
A. is left to the building or zoning official.
B. does not have to be completed.
C. is assigned to a fire inspector.
D. is given to the company officer that does company inspections for the area.

_____ **9.** During the plan review process, a good relationship should be maintained between the fire inspector and:
A. government regulatory agencies.
B. police agencies.
C. contractors.
D. Both A and C.

_____ **10.** The process of reviewing building plans and specifications to determine the safety characteristics of a proposed building refers to:
A. inspections.
B. citations.
C. plans review.
D. occupancy.

_____ **11.** When handling complaints, the fire inspector should:
A. only deal with the problem if it is an immediate problem.
B. take the complainant's name and phone number and forward the information.
C. take and process all pertinent information based on the complaint.
D. contact the officer of the engine company in the area of the complaint.

_____ **12.** When investigating a complaint, a fire inspector should anticipate that the occupant/owner may:
A. be cooperative and helpful.
B. be resistive.
C. deny the inspector access.
D. All of the above.

_____ **13.** When investigating a complaint, the fire inspector should document the situation. The reason for documenting the inspection is to:

A. provide evidence to the complainant that an inspection was made.

B. provide documentation for an appeals hearing.

C. help provide evidence to put the owner out of business.

D. All of the above.

_____ **14.** The key issue in allowing the fire code to be modified is whether the modification:

A. will look appealing to the public.

B. will impact the cost of the project.

C. will provide equal or greater protection.

D. is legal when compared to other buildings.

_____ **15.** The power given to the inspector to modify a fire code is determined by the:

A. National Fire Code.

B. International Conference of Building Codes.

C. National Fire Protection Association.

D. authority having jurisdiction.

_____ **16.** A/an ________________ should be in written form and include statistical data on the business, dates, times, phone numbers, violations, recommendations for corrections, and follow-up inspection date.

A. inspection report

B. code requirement

C. enforcement procedure

D. permit

_____ **17.** Which one of the following statements <u>is</u> <u>not</u> true?

A. Written reports serve not only as records of an inspection, but may be needed in legal proceedings.

B. Written correction orders should always include the inspector's personal opinions.

C. Without written evidence of an inspection, no proof exists that the inspector gave the owner notice of hazardous conditions or corrective measures to be taken.

D. The inspection report left with the owner should inform, analyze, and recommend.

_____ **18.** Which <u>is</u> <u>not</u> a concern when an inspector is completing an inspection report?

A. Proving the evidence

B. Presenting the facts

C. Recording the inspector's personal position

D. Justification of a recommendation

_____ **19.** In most cases, a formal inspection report will only be necessary:

A. when those inspections reported are on a standard checklist.

B. when site drawings have not been provided.

C. in cases of life-threatening hazards, major renovations, and/or an extensive list of violations.

D. when the inspector is unsure of a special requirement.

_____ **20.** Which one of the following resources **would not** be a good source for creating a checklist for the inspection of various occupancies?
A. NFPA - Fire Protection Handbook
B. NFPA - Conducting Fire Inspections, A Guide for Field Use
C. NFPA - Life Safety Code Handbook
D. NFPA - Field Incident Guide

_____ **21.** A well-constructed inspection form can be an invaluable tool for the fire inspector. Which of the following **should not be** included on the inspection form?
A. Inspector's name and title
B. Date and time of the inspection
C. The owner's reaction to the inspection
D. The business's name and the address of the building being inspected

_____ **22.** If an inspector discovers a code violation that **is not** on the inspection form, then he/she should:
A. ignore the violation because it is not on the inspection form.
B. document the violation on the form being used.
C. make a mental note to add the code violation to the form the next time the form is updated.
D. bring the violation to the owner's attention and give a verbal warning.

_____ **23.** By having a policies and procedures manual to follow, a fire inspector will have all of the following information **except**:
A. a knowledge of the limitations for fire inspectors.
B. how to handle complaints or questions.
C. how the appeals process works for the local authority.
D. how to interpret the fire code.

_____ **24.** Any person who performs fire inspections must have a thorough knowledge of all of the following **except**:
A. the statutes that designate them to perform inspections.
B. the appeal process.
C. the laws, codes, and ordinances that establish a fire inspector's duties and responsibilities.
D. their broad powers to mandate compliance.

_____ **25.** Using the floor area method, what is the occupant load factor for a structure 100 × 150 feet, one story, with two 36-inch wide exits for existing business occupancy?
A. 100
B. 150
C. 50
D. 200

_____ **26.** The single unit of exit width used to determine occupant load is _________________ inches.

A. 12
B. 18
C. 22
D. 24
E. 36

_____ **27.** The **minimum** number of exits usually required from any story or portion thereof shall be:

A. 4.
B. 2.
C. 5.
D. 3.

_____ **28.** You are asked to conduct an inspection at your local church. During the inspection you come across decorations which are used during the services and decide to conduct a field test to check for flammability. You should:

A. hold a piece of the material with pliers above a butane torch flame for two minutes.
B. apply a small flame from a common kitchen match to the material for twelve seconds.
C. apply a small flame from a match to the material for two minutes.
D. heat a sample of the material for one minute in a oven, at 350°, then apply a small flame for 30 seconds.

_____ **29.** The Steiner Tunnel test measures:

A. flame spread rating.
B. smoke developed.
C. fuel contributed.
D. All of the above.

_____ **30.** A chemical designed to retard ignition or the spread of fire when it is applied to material or another substance is defined as a:

A. fire proofer.
B. flame arrester.
C. fire rating.
D. fire retardant.

_____ **31.** A test designed to determine the surface burning characteristics of interior finishes is called a _________________ test.

A. carpet-pill
B. fire spread
C. Steiner Tunnel
D. flame spread

_____ **32.** A numerical rating assigned to a material based on the speed and extent to which flame travels over its surface is called a _______________ rating.
A. burning ember
B. fire resistance
C. flashover
D. flame spread

_____ **33.** The amount of time a material or assembly of materials will withstand a typical fire as measured on a standard time-temperature curve is called _______________ rating.
A. flame spread
B. fire resistance
C. fire curve
D. fire hazard

_____ **34.** Local codes or ordinances should specify driveway and entrance requirements for a facility based on:
A. previous agreements with other owners.
B. the number of apparatus that could respond to an emergency.
C. the largest fire apparatus that will be expected to respond to the occupancy.
D. the total number of apparatus used by the jurisdiction.

_____ **35.** _______________ or _______________ should specify driveway and entrance requirements that facilitate easy access off the street for the largest fire apparatus.
A. Local codes, ordinances
B. Model codes, Fire Inspector
C. Fire Marshal, Fire Inspector
D. Fire Marshal, NFPA

_____ **36.** An extremely important aspect of Life Safety is the _______________ to the site for fire apparatus and fire fighters.
A. clear width
B. accessibility
C. common path of travel
D. dead-end corridor

_____ **37.** A concern for firefighting personnel at any given location is the ability to place _______________ in a position where it can operate effectively during:
A. Incident Command, fireground training.
B. rehabilitation unit, fireground activities.
C. fire apparatus, emergency operations.
D. fireground activities, news media.

_____ **38.** Smoke control systems in new construction are designed to be used in conjunction with:
A. fire and smoke dampers.
B. fire doors and partitions.
C. curtain boards (draft curtains).
D. All of the above.

_____ **39.** In performance-based design, the acceptance test criteria, methods of evaluating data, and approval procedures are developed by the:

A. architect and contractor.

B. building owner.

C. authority having jurisdiction.

D. planning commission.

E. All of the above.

_____ **40.** Construction classifications are based on:

A. regional locations.

B. authority having jurisdiction.

C. the materials that are contained in the building.

D. the materials used in construction of the building.

_____ **41.** A smoke-proof enclosure in a three story building must be enclosed from its highest point to its lowest point by a ________________-hour fire rating.

A. one

B. two

C. three

D. four

_____ **42.** Dry chemical fixed fire extinguishing systems **are not** recommended for use on:

A. flammable liquids.

B. flammable gases.

C. delicate electronic equipment.

D. paper storage.

_____ **43.** Which of the following are types of fixed fire extinguishing systems?

A. Automatic sprinkler systems, foam systems, carbon dioxide systems, halogenated agent systems, and chemical systems

B. Automatic sprinkler systems, factory mutual systems, and U.L. listed systems

C. Wet sprinkler systems, dry sprinkler systems, deluge systems, fire mutual systems, and hazardous substance systems

D. Water sprinkler systems, air sprinkler systems, flooding sprinkler systems, and chemical sprinkler systems

_____ **44.** Prior to pumping, the amount of pressure that is expected to be available from a hydrant is called normal ________________ pressure.

A. atmospheric

B. operating

C. residual

D. grid

_____ **45.** Which of the following <u>is</u> <u>not</u> a primary water supply source for an automatic sprinkler system?
A. Public water supply system
B. Gravity tank
C. Pressure tank
D. Fire department water tanker truck

_____ **46.** _________________ will completely line the interior of a pipe and will gradually cause a restricted diameter.
A. Lead deposits
B. Sedimentation
C. Encrustations
D. Mud deposits

_____ **47.** _________________ deposits consist of mud, clay, and dead organisms.
A. Valve
B. Sedimentary
C. Tuberculation
D. Encrustation

_____ **48.** The upright-type of sprinkler head:
A. may only be used in a wet-pipe configuration.
B. protrudes downward from the exposed pipe.
C. cannot be inverted for use in the pendant-type sprinkler head.
D. can be installed in a side wall application.

_____ **49.** The NFPA standard which requires that fire extinguishers be thoroughly inspected is:
A. NFPA 25.
B. NFPA 14.
C. NFPA 13.
D. NFPA 10.

_____ **50.** All maintenance procedures should include a thorough examination of the three basic parts of a portable extinguisher:
A. mechanical parts, extinguishing agent, and expelling means.
B. mechanical parts, color, and expelling means.
C. gauge, size, and color.
D. extinguishing agent, nozzle, and gauge.

_____ **51.** A network of intermediate-sized pipe that reinforces the overall grid system by forming loops that interlock primary feeders <u>best</u> defines:
A. primary loop.
B. secondary feeders.
C. distributors.
D. grid network.

_____ **52.** Large pipes which carry large quantities of water to various points along the water supply system for distribution to smaller mains **best** defines:
A. primary feeders.
B. secondary feeders.
C. distributors.
D. grid network.

_____ **53.** Today, 8" pipe is becoming the minimum size pipe in a water distribution system because:
A. of its higher flow capacity advantages over smaller diameter pipe.
B. all pipe would then be the same diameter.
C. the pipe size determines the flow pressures needed to design a reliable system.
D. new code requirements are phasing out small diameter pipe.

_____ **54.** The standard hydrostatic test for all piping in a wet pipe systems is **not less than** ______________ for ______________.
A. 50 psi, 1 hour
B. 200 psi, 2 hours
C. 150 psi, 20 minutes
D. 100 psi, 1-1/2 hour

_____ **55.** Which of the following **would not** be considered an obsolete portable fire extinguisher?
A. Soda-acid
B. Water cartridge operated
C. Vaporization liquid
D. Stored pressure

_____ **56.** Exit stairwell pressurization is a method of:
A. signal notification.
B. fire alarm systems.
C. fire control.
D. smoke control.

_____ **57.** After conducting a trip test on a dry-pipe system, water and air pressure gauges should:
A. be the same as before the test.
B. indicate equal pressure.
C. not read greater than 110 psi.
D. return to zero.

_____ **58.** When inspecting any dry-pipe sprinkler system, the air pressure gauge should read approximately:
A. the same as the water pressure.
B. the same as recorded on the previous test.
C. three times atmospheric pressure.
D. half of the gauge's full-scale pressure.

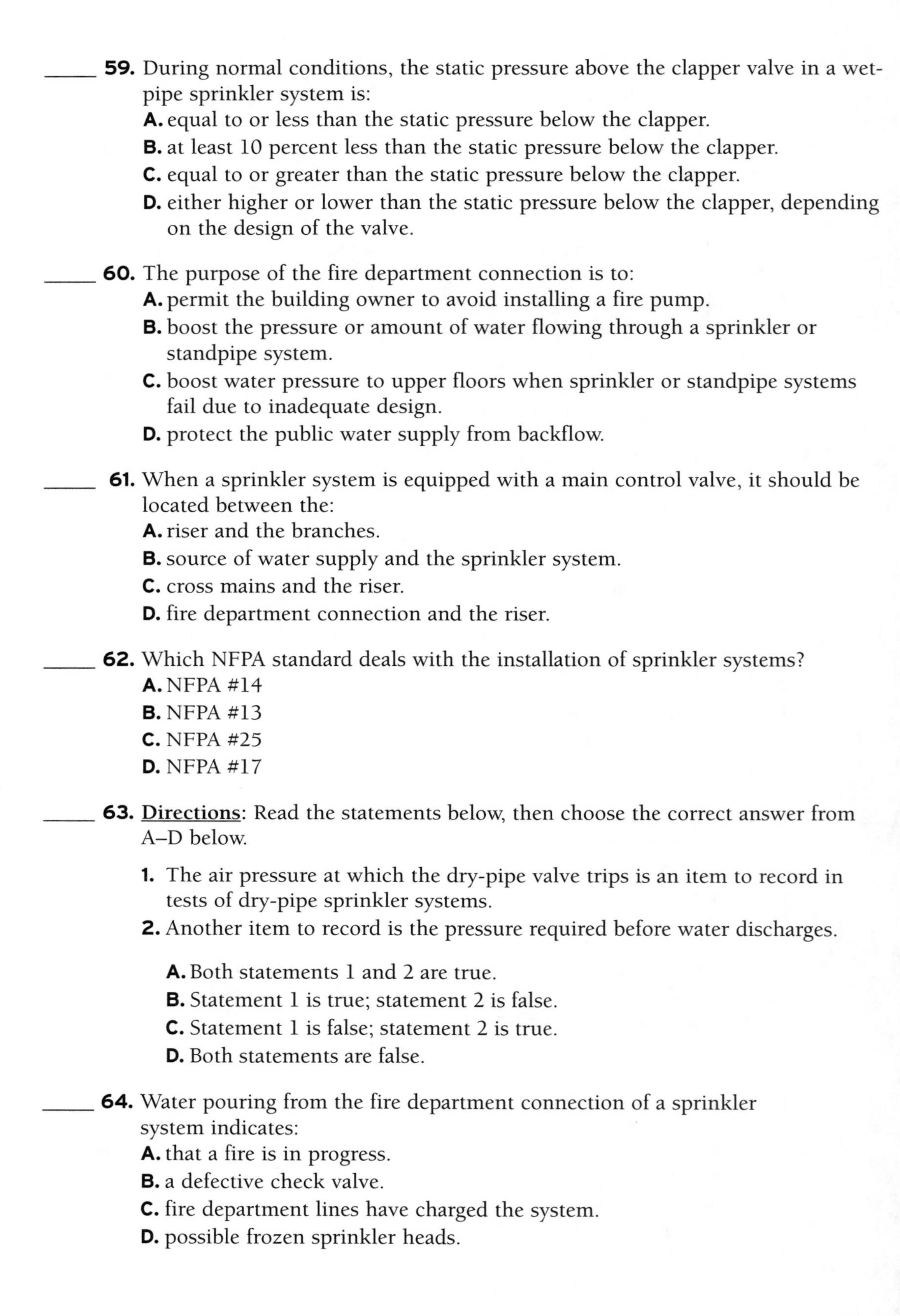

_____ **59.** During normal conditions, the static pressure above the clapper valve in a wet-pipe sprinkler system is:
A. equal to or less than the static pressure below the clapper.
B. at least 10 percent less than the static pressure below the clapper.
C. equal to or greater than the static pressure below the clapper.
D. either higher or lower than the static pressure below the clapper, depending on the design of the valve.

_____ **60.** The purpose of the fire department connection is to:
A. permit the building owner to avoid installing a fire pump.
B. boost the pressure or amount of water flowing through a sprinkler or standpipe system.
C. boost water pressure to upper floors when sprinkler or standpipe systems fail due to inadequate design.
D. protect the public water supply from backflow.

_____ **61.** When a sprinkler system is equipped with a main control valve, it should be located between the:
A. riser and the branches.
B. source of water supply and the sprinkler system.
C. cross mains and the riser.
D. fire department connection and the riser.

_____ **62.** Which NFPA standard deals with the installation of sprinkler systems?
A. NFPA #14
B. NFPA #13
C. NFPA #25
D. NFPA #17

_____ **63.** Directions: Read the statements below, then choose the correct answer from A–D below.

1. The air pressure at which the dry-pipe valve trips is an item to record in tests of dry-pipe sprinkler systems.
2. Another item to record is the pressure required before water discharges.

A. Both statements 1 and 2 are true.
B. Statement 1 is true; statement 2 is false.
C. Statement 1 is false; statement 2 is true.
D. Both statements are false.

_____ **64.** Water pouring from the fire department connection of a sprinkler system indicates:
A. that a fire is in progress.
B. a defective check valve.
C. fire department lines have charged the system.
D. possible frozen sprinkler heads.

_____ **65.** The temperature range of a sprinkler head color-coded red is _________________ degrees F.
A. 175 to 225
B. 250 to 300
C. 325 to 375
D. 400 to 475

_____ **66.** Within a fire protection system, an electronic circuit that monitors the system's readiness and transmits a signal when there is a problem with the system is a:
A. voltage unit.
B. mechanical circuit.
C. voltage circuit.
D. supervisory circuit.

_____ **67.** Which of the following are part of a means of egress?
A. Corridor, stairs, alley
B. Overhead door, non-rated corridor, alley
C. Through kitchen, corridor, street
D. Through mechanical room, fenced in yard, street

_____ **68.** When illumination of exits is required, it must be continuous during occupancy and floors must be illuminated at not less than 1 foot candle (10.8 Lumens DX) measured at the:
A. floor.
B. ceiling.
C. wall.
D. ceiling and floor.

_____ **69.** When determining the height of a building, the jurisdiction's codes look at all of the following factors <u>except</u>:
A. lowest point to consider.
B. total number of occupants.
C. highest point to consider.
D. automatic sprinkler protection.

_____ **70.** In addition to determining how many people may occupy a building, an inspector must determine the number of exits. This is part of the:
A. load capacity.
B. means of egress.
C. means of access.
D. means of exit.

_____ **71.** The four main views of working drawings are:
A. floor plan, mechanical, electrical, and fire protection.
B. fire protection, elevation, sectional, and floor plan.
C. plan, elevation, sectional, and detailed.
D. mechanical, sectional, elevation, and site plan.

_____ **72.** Cooking equipment must have a clearance of at least _________________ inches to any combustible material.
A. 6
B. 10
C. 12
D. 18

_____ **73.** A common hazard associated with central heating appliances, unit heaters, and room heaters is:
A. temperature limit controls.
B. proximity of installation to combustibles.
C. asphyxiation due to inadequate make-up air.
D. the explosion potential.

_____ **74.** Loading and unloading stations for Class I liquids should be no closer than _________________ feet from storage tanks, property lines, or adjacent buildings?
A. 25
B. 10
C. 15
D. 50

_____ **75.** In doing a plan review, which NFPA standard would apply to the installation of an air conditioning and ventilating system?
A. NFPA 89
B. NFPA 96
C. NFPA 90A
D. NFPA 97

_____ **76.** School fire drills should be conducted:
A. at different times of the day.
B. only when it is convenient for the teachers.
C. at the beginning of classes.
D. at the discretion of the local fire department.

_____ **77.** The facilities that **do not** warrant total evacuation every time the fire alarm sounds are:
A. daycare centers, schools, and hotels.
B. hospitals and nursing homes.
C. hospitals, schools, and daycare centers.
D. hotels, motels, and nursing homes.

_____ **78.** Tanks designed to hold large quantities of low pressure gas are built according to standards set forth by the:
A. U.S. Department of Transportation.
B. Boiler and Pressure Vessel Code.
C. American Petroleum Institute (API).
D. American National Standards Institute (ANSI).

_____ **79.** Portable tanks with capacity greater than _______________ gallons should be treated according to the same standards as fixed tanks?
A. 60
B. 120
C. 240
D. 660

_____ **80.** Flammable or combustible liquid containers stored in warehouses or buildings must be arranged to provide access aisles of **approximately** _______________ in width.
A. 3 feet
B. 6 feet
C. 9 feet
D. 12 feet

_____ **81.** Above ground atmospheric tanks for flammable or combustible liquids are designed for pressures of _______________ psi.
A. 0 to 0.5
B. 0.5 to 15
C. 15 to 20
D. 10 to 15

_____ **82.** Flammable gases are stored in a variety of _______________ and _______________.
A. cylinders, tanks
B. drums, barrels
C. bottles, pipelines
D. pipelines, cylinders

_____ **83.** The temperature at which the vapor pressure of a liquid is equal to the external pressure applied to it is the:
A. fire point.
B. melting point.
C. flash point.
D. boiling point.

_____ **84.** The **best** source of information on a specific hazardous material from the manufacturer is the:
A. Material Safety Data Sheet (MSDS).
B. Coast Guard Hazard Response Information System Manual (CHRIS Manual).
C. Bureau of Alcohol, Tobacco, and Firearms (ATF).
D. American Association of Rail Roads (AARR).

_____ **85.** Buildings or structures used primarily for securing and sheltering goods, merchandise, products, vehicles, or animals are classified as _______________ occupancies.
A. storage
B. mercantile
C. business
D. warehouse

_____ **86.** According to NFPA 101, Life Safety Code, there are ________________ occupancy classifications.
A. 3
B. 9
C. 17
D. 33

_____ **87.** The classification given to a particular building based on the materials and methods used to construct and the ability of those materials to resist the effects of a fire situation is the ________________ classification.
A. construction
B. occupancy
C. flame spread
D. fire hazard

_____ **88.** A sketch depicting the general arrangement of the property in reference to streets, adjacent properties, and other important features is known as a(n):
A. site plan.
B. sectional view.
C. floor plan.
D. aerial view.

_____ **89.** When exits serve more than one floor, an important element in determining the exit capacity is the:
A. maximum square feet per floor.
B. overall height allowable.
C. maximum occupant load per floor.
D. All of the above.

_____ **90.** A dance floor located in a drinking establishment measures 18 feet by 24 feet. Calculate the occupant load using the floor area method using 7 square feet per person.
A. 9
B. 15
C. 29
D. 62
E. 69

_____ **91.** When a fire hydrant receives water from two or more directions, it is said to have ________________ feed.
A. compound
B. circulating
C. compensating
D. distributed

_____ **92.** Valve spacing for water systems in high value districts should be no more than _______________ feet.
A. 300
B. 500
C. 750
D. 1,000

_____ **93.** Which of the following **would** **not** normally be part of a grid system?
A. Primary feeders
B. Secondary feeders
C. Inter-connecting distributors
D. Dead-end water mains

_____ **94.** When computing available fire flow in a specific area, the required residual pressure is _______________ psi.
A. 5
B. 10
C. 15
D. 20

_____ **95.** When an opening in a three-hour fire-rated wall needs a fire-rated door, the requirement should be for:
A. 2 two-hour doors on one side.
B. three-hour doors, one on each side.
C. 2 two-hour doors, one on each side.
D. 1 three-hour door.

_____ **96.** Building construction techniques and materials are tested under two NFPA standards. What are they?
A. NFPA 255 and 257
B. NFPA 251 and 255
C. NFPA 252 and 257
D. NFPA 257 and 251

_____ **97.** According to NFPA 220, Standard on Types of Building Construction, Type II construction is similar to Type I construction **except** that the degree of fire resistance is:
A. lower.
B. higher.
C. fire rated.
D. approved by NFPA.

_____ **98.** The roof covering that provides the **best** fire retardant properties is:
A. Class A.
B. Class B.
C. Class C.
D. Class D.

_____ **99.** According to NFPA 256, Methods of Fire Tests of Roof Coverings, the roof covering that provides the least fire retardant properties is:

A. Class A.
B. Class B.
C. Class C.
D. Class D.

_____ **100.** Heavy timber construction is allowed for interior structural members in _______________ construction.

A. Type I
B. Type II
C. Type III
D. Type IV
E. Type V

Did you score higher than 80 percent on Examination II-3? Circle Yes or No in ink.

Feedback Step

What do we do with your "yes" and "no" answers through the NFPA Standard 1031 examination preparation process? First, return to any "no" response. Go back to the highlighted answers for those examination items missed, and read and study the paragraph preceding the location of the answer as well as the paragraph *following* the paragraph where the answer is located. This will expand your knowledge base for the missed question, put it in a broader context, and improve associative learning. Remember, we are trying to develop mastery of the required knowledge. Scoring 80 percent on an examination is good but it is not mastery performance. To come out in the top of your group, you must score much higher than 80 percent on your training, promotion, or certification examination.

Phases III and IV focus on getting you ready for the examination process by recommending activities that have a positive impact on the emotional and physical part of examination preparation. By evaluating your own progress through SAEP, you have determined that you have a high level of knowledge. Taking an examination for training, promotion, or certification is a competitive event. Just as in sports, total preparation is vitally important. Now you need to get all the elements of good preparation in place so that your next examination experience will be your best ever. Before moving on to Phase III, review "Summary of Key Rules for Taking an Examination" and "Summary of Helpful Hints."

Summary of Key Rules for Taking an Examination

Rule 1—Examination preparation is not easy. Preparation is 95 percent perspiration and 5 percent inspiration.

Rule 2—Follow the steps very carefully. Do not try to reinvent or shortcut the system. It really works just as it was designed to!

Rule 3—Mark with an "X" any examination items for which you guessed the answer. To obtain the maximum return on effort, research any answer that you guessed even if you guessed correctly. Find the correct answer, highlight it, and then read the entire paragraph that contains the answer. Be honest and mark all questions on which you guessed. Some examinations have a correction for guessing built into the scoring process. The correction for guessing can reduce your final examination score. If you are guessing, you are not mastering the material.

Rule 4—Read questions twice if you have any misunderstanding and especially if the question contains complex directions or activities.

Rule 5—If you want someone to perform effectively and efficiently on the job, the training and testing program must be aligned to achieve this result.

Rule 6—When preparing examination items for job-specific requirements, the writer must be a subject matter expert with current experience at the level that the technical information is applied.

Rule 7—Good luck = good preparation.

Summary of Helpful Hints

Helpful Hint—Your first impression is often the best. More than 41 percent of changed answers during our SAEP field test were changed from a right answer to a wrong answer. Another 33 percent changed their answer from a wrong answer to another wrong answer. Only 26 percent of answers were changed from wrong to right. In fact, three participants did not make a perfect score of 100 percent because they changed one right answer to a wrong one! Think twice before you change your answer. The odds are not in your favor.

Helpful Hint—Researching correct answers is one of the most important activities in SAEP. Locate the correct answer for all missed examination items. Highlight the correct answer. Then read the entire paragraph containing the answer. This will put the answer in context for you and provide important learning by association.

Helpful Hint—Work through all missed examination items using the same technique. Reading the entire paragraph improves retention of the information and helps you develop an association with the material and learn the correct answers. This step may sound simple. A major finding during the development and field testing of SAEP was that you learn from your mistakes.

Helpful Hint—Follow the steps carefully to realize the best return on effort. Would you consider investing your money in a venture without some chance of return on that investment? Examination preparation is no different. You are investing time and expecting a significant return for that time. If, indeed, time is money then you are investing money and are due a return on that investment.

Helpful Hint—Try to determine why you selected the wrong answer. Usually something influenced your selection. Focus on the difference between your wrong answer and the correct answer. Carefully read and study the entire paragraph containing the correct answer. Highlight the answer just as you did for the other examinations.

Helpful Hint—Studying the correct answers for missed items is a critical step in return on effort! The focus of attention is broadened and new knowledge is often gained by expanding association and contextual learning. During our research and field test, self-study during this step of SAEP resulted in gains of 17 points from the first examination administered to the third examination. An increase in your score of 17 points can move you from the lower middle to the top of the list of persons taking a training, promotion, or certification examination. This is a competitive edge and a prime example of return on effort in action. Remember: Maximum effort = Maximum results!

PHASE III

How Examination Developers Think: Getting Inside Their Heads

Now that you've finished the examination practice, this additional information will assist you in understanding and applying examination-taking skills. Developing your knowledge of how examination professionals think and prepare examinations is not cheating. Indeed, most serious examination takers have spent many hours reviewing various examinations to gain an insight into the technology. It is a demanding technology when used properly. You probably already know this if you have prepared examination items and administered them in your fire department. Phase III will not cover all the ways and means of examination-item writing. Examination-item writers use many techniques—far too many to cover in this book. The focus here is on key techniques that will help you achieve a better score on your examination.

How are examination items derived?

Professional examination item writers use three basic techniques to derive examination items from text or technical reference materials: verbatim, deduction, and induction.

The most common technique is to take examination items verbatim from materials in the reference list. This technique doesn't work well for mastering information because it encourages rote learning or memorizing the material. The results of this type of learning are not long lasting, nor are they appropriate for learning and retaining the critical knowledge you must have for on-the-job performance. The SAEP process doesn't create the majority of examination questions for NFPA Standard 1031 using the verbatim technique. Professional examination-item writers tend to use verbatim testing at the very basic level of job classifications. A first responder, for instance, is expected to learn many basic facts. At this level, verbatim examination items can be justified.

In the higher ranks of the Fire and Emergency Medical Service profession, other methods are more beneficial and productive for mastering higher cognitive knowledge and skills. Examination-item development at the higher cognitive levels of an occupation, such as Fire Officer, will rely on other means. The most important consideration at such cognitive levels is using deduction as the basis for examination items. This technique requires logic and analytical skills and often requires the examination taker to read materials several times to answer the examination item. It is not a matter of reproducing the information that results in a verbatim answer. At the first responder level, most activities are carefully supervised by a more experienced technician or company officer. The responder is expected to closely follow commands and is encouraged not to use deductive reasoning that can lead to "freelance" responder tactics. As the individual progresses to the Fire Inspector II level and gains experience, deductive reasoning and inference skills are developed and applied. Most deductive reasoning and inference skills are related to personal safety and the safety of those on the scene. Most sizeup strategies are developed and passed from the officers on the scene to the first responders.

Rule 5

If you want someone to perform effectively and efficiently on the job, the training and testing program must be aligned to achieve this result.

Rule 5 is paramount for first responders. Effective and efficient first responders are able to receive fireground commands, follow instructions, and perform their tasks safely and as rapidly as they can. There are limited opportunities for first responders to do much else, as they are the first line of action at the emergency scene.

An example of deductive reasoning: an incident call is received from the Telecommunicator stating that an infant has a high temperature and is convulsing. Just this amount of information should cause the first responder to immediately plan the response, conduct sizeup activity, and review infant care procedures en route. Some of these deductive responses will focus on the infant's age, past medical history, location, and access, among many other possible deductions. If you have an EMT or paramedic background, a list of several items could be deduced that would expedite an efficient and effective response to the incident.

You can probably think of many first responder tasks and circumstances that rely on deductive reasoning. The more experience you gain on the fireground as a firefighter, and the more frequently you practice deductive reasoning and inference from emergency data, the more efficient and effective you will become in emergency situations, whether you are ventilating a roof or attending to the emergency needs of an infant.

Legendary football coach Vince Lombardi was once asked about the precision performance of his offensive and defensive teams. The comment was made that Lombardi must spend a lot of time on the practice field to get those results. He was quoted as saying, "Practice doesn't make perfect; only perfect practice makes perfect." This is exactly what is required to be an outstanding examination taker. Most people do not perfectly practice examination-taking skills.

A third examination-item development technique relies on inference—that is, implied answers. Once again, the examination-item writer doesn't rely on verbatim techniques. Inference requires contrasting, comparing, analyzing, evaluating, and other high-level cognitive skills. Tables, charts, graphs, and other instruments for presenting data provide excellent means for deriving inference-based examination items. Implied answers are logic based. They rely on your ability to use logical processes or a series of facts to arrive at a plausible answer. For example, recent data supplied by the NFPA stated that heart attacks remain the leading cause of death for fire service personnel. Other data stated that strains and sprains remain the leading cause of injuries on the job. Inferences can be made from this relatively simple statement of information. A Safety Officer can apply these data to his or her own personnel and use the information as a basis for checking on personnel, conducting surveys, reviewing accident records, and comparing the results with actual experience. Is your Fire Department doing better or worse in these important health issues? Are your fire and emergency service personnel getting the right amounts and kinds of exercise? Are you diligent in keeping the station and fireground free from the activities that may lead to strains and sprains? The basic inference here is that your fire department is similar or different in some ways from the generalized data.

If you find it difficult to find an answer to an examination item, it may be that the question is measuring your ability to deduce and make inferences or to get the implied answer from the technical materials.

How are examination items written and validated?

Once the appropriate information is identified and the technique for writing an examination item selected, the professional will prepare a draft. The draft examination item is then referenced to specific technical information such as a textbook, manufacturer's manual, or other related technical information. If the information is derived from a job-based requirement, then it should also be validated by job incumbents (i.e., those who are actually performing in the occupation at the specific level of the required knowledge).

Rule 6

When preparing examination items for job-specific requirements, the writer must be a subject matter expert with current experience at the level where the technical information is applied.

Rule 6 ensures that the examination item has a basic level of job-content validity. The final level of job-content validity is determined by committees or surveys of job incumbents who certify the information to be current and required on the job. The information must be categorized as "need to know" or "must know" to be considered job relevant. The technical information must be accurate. Because subject matter experts do need basic training in examination-item writing, it is recommended that a professional in examination technology be part of the review process so that basic rules and guidelines of the industry are followed.

Finally, the examination items must be field tested. Once this testing is complete, statistical and analytical tools are available to help revise and improve the examination items. These techniques and tools go well beyond the scope of this *Exam Prep* book. Professionals are available to conduct these data analyses, and their services should be used.

Good Practices in Examination-Item and Examination Development

The most reliable examinations are objective. That is, they have only one answer that is accepted by members of the occupation. This objective quality permits fair and equitable examinations. The most popular objective examination items are multiple choice, true/false, matching, and completion (fill in the blanks).

Ten rules govern the development of valid and reliable job-relevant examinations in the Fire and Emergency Medical Service:

1. They do not contain trick questions.
2. They are short and easy to read, using language and terms appropriate to the target examination population.
3. They are supported with technical references, validation information, and data on their difficulty, discrimination, and other item-analysis statistics.
4. They are formatted to meet recognized testing standards and examples.
5. They focus on the "need to know" and "must know" aspects of the job.
6. They are fair and objective.
7. They are not based on obscure and trivial knowledge and skills.
8. They can be easily defended in terms of job-content requirements.
9. They meet national and other professional job qualification standards.
10. They demonstrate their usefulness as part of a comprehensive testing program, including written, oral, and performance examination items.

The primary challenges in ensuring job-relevant examinations relate to their currency and validity. Careful recording of data, technical reference sources, and examination writer qualifications are important. Examinations that affect someone's ability to be promoted, certified, or licensed, as well as to complete training that leads to a job, have exacting requirements both in published documents and in the laws of the land.

Three Myths of Examination Construction

1. **Myth:** If in doubt about the answer, select the longest answer in a multiple-choice examination item.
   Reality: Professional examination-item writers use short answers as correct ones at an equal or higher percentage than longer answers. Remember, there are usually choices A–D. That leaves three possibilities for the correct answer other than the longest one. Statistically speaking, the longest answer is less likely to be correct.
2. **Myth:** If in doubt about the answer in a multiple-choice examination item, select "C".
   Reality: Computer technology and examination-item banking permit multiple versions of examinations to be simultaneously developed. This is typically achieved by moving the correct answer to different locations (for example, version 1 will have the correct answer in the "C" position, version 2 in the "D" position, and so forth).
3. **Myth:** Watch for mismatches between singular examination-item stems and plural choices in the A–D answers, or vice versa.
   Reality: Most computer-based programs have spelling and grammar checking utilities. If this mistake occurs, an editing error is the probable cause and usually has nothing to do with detecting the right answer.

Some Things That Work

1. Two to three days before your examination, review the examination items you missed in SAEP. Read those highlighted answers and the entire paragraph one more time.
2. During the examination, carefully read the examination item twice. Once you have selected your answer, read the examination item and answer together. This technique can key information that has been studied during your examination preparation activities.
3. Apply what you learned in SAEP. Eliminate as many distracters as possible to improve the probability of getting the correct answer.
4. Pace yourself. Know how much time you have to take the examination. If an examination item is requiring too much time, write its number down and continue with the next examination item. Often, an examination item later in the examination will trigger your memory and make the examination item easier to answer. (For a time pacing strategy see the examination pacing table at the end of Phase IV.)
5. Don't panic if you don't know some examination items. Leave them to answer later. The most important thing is to finish the examination because you may know several examination items at the end of the examination.
6. As time runs out for taking the examination, don't panic. Concentrate on answering those difficult examination items that you skipped.

7. Double-check your answer sheets to make sure you have not accidentally left an answer blank.
8. Once you complete the examination, return to the difficult examination items. Often, completing other examination items will cause you to remember or associate those answers with the difficult examination-item answers. The longer the examination, the more likely you will be to gather the information needed to answer the difficult examination items.

Many other helpful hints can be used. If you want to research other materials on how to take examinations and raise your final score, visit your local library, bookstore, or the Web for additional resources. The main reason we developed SAEP is to provide practice and help you develop examination-taking skills that you can use throughout your life.

PHASE IV

The Basics of Mental and Physical Preparation

Mental Preparation: I Can Get My Head Ready!

The two most common mental blocks to examination taking are examination anxiety and fear of failure. In the Fire and Emergency Medical Service, these feelings can cause significant performance barriers. Severe conditions may require some professional psychological assistance, which is beyond the scope of this book.

The root cause of examination anxiety and fear of failure is often a lack of self-confidence. SAEP was designed to help improve self-confidence by providing evidence of mastery. Look at your scores as you progress through Phase I or Phase II. Review your Personal Progress Plotter (found at the end of the Introduction section). It will help you gain confidence in your knowledge of NFPA Standard 1031. Look at your Personal Progress Plotter the day before your scheduled examination and experience renewed confidence.

Let's examine the meaning of anxiety. Knowing what it is will help you deal with it at examination time. According to *Webster's Dictionary*, anxiety is "uneasiness and distress about future uncertainties." Many of us have real anxiety about taking examinations, and it is a natural response for some. This experience is often prefaced by several questions: Am I ready for this? Do I have a good idea of what is going to be on the examination? Will I make the lowest score? Will John Doe score higher than me?

These questions and concerns are normal. Remember that hundreds of people have gone through SAEP with an average increase of 17 points in their scores. The preparation process will help you maintain confidence. Once again, check the evidence in your Personal Progress Plotter to see what you have accomplished.

Fear, according to *Webster's Dictionary*, is "alarm and agitation caused by the expectation or realization of danger." Let's deal with this normal reaction to examinations. First, analyze the degree of fear you may be experiencing several days before the examination date. Then, focus on the positive experiences you had as you finished SAEP. Putting your fear into perspective by using positives to eliminate or minimize it is a very important examination-taking skill. The more you focus on your positive accomplishments in mastering the materials, the less fear you will experience.

If your fear and anxiety persist, even after you take steps to build your confidence, you may want to get some professional assistance. Do it now! Don't wait until the week before the examination. There may be real issues that the professional can help you deal with to overcome these feelings. Hypnosis and other forms of treatment have been found to be very helpful. Consult with a professional expert.

Physical Preparation: Am I Really Ready?

Physical preparation is the element that is probably most ignored in examination preparation. In the Fire and Emergency Medical Service, examinations are often given at locations away from home. If this is the case, you need to be especially careful of key physical concerns.

In general, following these helpful hints will help you concentrate, enhance your examination performance, and add points to your score.

1. Do not "cram" for the examination. This strategy was found to be first in importance during our field testing of SAEP. Cramming results in examination anxiety, adds to confusion, and tends to lessen the effectiveness of examination-taking skills you already possess. Avoid cramming!
2. Get a normal night's rest. It may even be wise to take a day off before the examination to rest. Do not schedule an all-night shift immediately before your examination.
3. Avoid taking excessive stimulants or medications that might inhibit your thinking processes. Eat at least three well-balanced meals before the day of the examination. It is a good practice to carry a nutritionally balanced energy bar (not candy) and a bottle of water into the examination area. Examination anxiety and fear can cause a dry mouth, which can lead to further aggravation. Nibbling on the energy bar also has a settling effect and supplies some "brain food."
4. If the examination is at an out-of-town location, take the following steps:
    - Avoid a "night out with friends." Lack of rest, partying, and fatigue are major examination performance killers.
    - Check your room carefully. Eliminate things that might aggravate you, interfere with your rest, or cause any discomfort. If the mattress is not good, pillows are horrible, or the room has an unpleasant odor, change rooms or even hotels.
    - Wake up in plenty of time to take a relaxing shower or soaking bath. Do not put yourself in a "rush" mode. Things should be carefully planned so that you arrive at the examination site ahead of time, calm, and collected.
5. Listen to the examination proctor. The proctor typically has rules that you must follow. Important instructions and directions are usually given. Ask clarifying questions immediately and listen to the responses to questions raised by the other examination takers. Most examination environments are carefully controlled and may not permit questions you raise that are covered in the proctor's comments or deal with the technical content in the examination itself. Be attentive, focus, and succeed.
6. Remain calm and breathe. Pace yourself. Apply your examination-taking skills learned during SAEP.
7. Remember the analogy of an examination as a competitive event. If you want the competitive edge, then carefully follow all phases of SAEP. This process has yielded outstanding results in the past and can do the same for you.

Time Management During Examinations

The examination pacing table will help you pace yourself during an examination. You should get familiar with the table and be able to construct your own when you are in the examination room and getting ready to start the examination process. It will take a few minutes but it will make a tremendous contribution to your time management during the examination.

Here is how the table works. First you divide the examination time into six equal parts. If you have 3½ hours (210 minutes) for the examination, then each of the six parts contains 35 minutes: 210 ÷ 6 = 35 minutes. Now divide the number of examination items by 5. For example, if the examination has 150 examination items, 150 ÷ 5 = 30. With the math done, you can set up a table that tells you approximately how many examination items you should answer in 35 minutes (the equal time divisions). You should be on or near examination item 30 at the end of the first 35 minutes, and so forth. Notice that we divided the examination items by 5 and the time by 6. This extra time block of 35 minutes is used to double-check your answer sheet, focus on difficult questions, and calm your nerves. This technique will work wonders for your stress level and, yes, will improve your examination score.

Examination Pacing Table (150 and 100 Examination Items)

Time for Examination	Minutes for Six Equal Time Parts	Number of Examination Items	Examination Items per Time Part	Time for Examination Review
210 minutes (3.5 Hours)	35 minutes	150	30 (number of examination items to be answered)	35 minutes (chilling and double-checking examination)
150 minutes (2.5 Hours)	25 minutes	100	20 (number of examination items to be answered)	25 minutes (chilling and double-checking examination)

The examination pacing table can be adjusted by modifying the time/examination-item variables, as either may change in the real examination environment. For instance, if the time available changes, adjust the ratio of time available to answer the examination items in each of the five time blocks. If examination-item numbers increase or decrease, adjust the number of examination items to be answered in the time blocks.

Take some precautions when using this time management strategy:

1. Don't panic if you run a few minutes behind in each time block. The time management strategy should not stress you while using it. Most people tend to pick up their pace as they move into the examination.
2. During the examination, carefully mark or note examination items that you need to return to during your review time block. This will help you expedite your examination completion check.
3. Don't be afraid to ask for more time to complete your examination. In most cases, the time limit is flexible (or should be).
4. Double-check your answer sheet to make sure that you didn't leave blank responses and that you didn't double-mark answers. Double-markings are most often counted as wrong answers. Ensure that any erasures are made cleanly. Caution: Make sure when you change your answer that you really want to do so. Odds are not in your favor unless something on the examination really influenced the change.

APPENDIX A

Examination I-1 Answer Key

Directions

Follow these steps carefully for completing the feedback part of the Systematic Approach to Examination Preparation (SAEP):

1. After entering your scores, look up the answers for the examination items you missed as well as those you guessed, even if you guessed correctly. If you are guessing, it means the answer isn't perfectly clear. This process will make you as knowledgeable as possible.
2. Enter the number of missed and guessed examination items in the blank on your Personal Progress Plotter.
3. Highlight the answer in the reference materials, and then read the paragraph preceding and the paragraph following the one in which the correct answer is located. Enter the paragraph number and page number next to the guessed or missed examination item on your examination. Count any part of a paragraph at the beginning of the page as one paragraph until you reach the paragraph containing your highlighted answer. This step will help you locate and review your missed and guessed examination items later in the process. This step is essential to learning the material in context and by association. These learning techniques (context/association) are the very backbone of the SAEP approach.
4. Once you have completed the feedback part, you may proceed to the next examination.

1. Reference: NFPA 1031, 4.2.1 and 4.2.1(A)(B)
   IFSTA, *Fire Inspection and Code Enforcement,* 6th Edition, 1st Printing, page 13.
   Answer: D

2. Reference: NFPA 1031, 4.2.1 and 4.2.1(A)(B)
   IFSTA, *Fire Inspection and Code Enforcement,* 6th Edition, 1st Printing, page 13.
   Answer: C

3. Reference: NFPA 1031, 4.2.1, 4.2.1(A)(B), 4.3.15, and 4.3.15(A)(B)
   IFSTA, *Fire Inspection and Code Enforcement,* 6th Edition, 2nd Printing, pages 40–41.
   Answer: A

4. Reference: NFPA 1031, 4.2.1, 4.2.1(A), 4.2.2 and 4.2.2(A)
   IFSTA, *Fire Inspection and Code Enforcement,* 6th Edition, 2nd Printing, page 37.
   Answer: D

5. Reference: NFPA 1031, 4.2.2 and 4.2.2(A)(B)
   IFSTA, *Fire Inspection and Code Enforcement,* 6th Edition, 1st Printing, pages 19–20.
   Answer: B

6. Reference: NFPA 1031, 4.2.3 and 4.2.3(A)(B)
IFSTA, *Fire Inspection and Code Enforcement,* 6th Edition, 2nd Printing, page 290.
Answer: C

7. Reference: NFPA 1031, 4.2.3 and 4.2.3(A)(B)
IFSTA, *Fire Inspection and Code Enforcement,* 6th Edition, 2nd Printing, page 289.
Answer: B

8. Reference: NFPA 1031, 4.2.4 and 4.2.4(A)(B)
IFSTA, *Fire Inspection and Code Enforcement,* 6th Edition, 1st Printing, pages 14–15.
Answer: A

9. Reference: NFPA 1031, 4.2.4 and 4.2.4(A)(B)
IFSTA, *Fire Inspection and Code Enforcement,* 6th Edition, 1st Printing, pages 14–15.
Answer: D

10. Reference: NFPA 1031, 4.2.5, 4.2.5(A)(B), 4.3.15, and 4.3.15(A)(B)
IFSTA, *Fire Inspection and Code Enforcement,* 6th Edition, 1st Printing, page 12.
Answer: C

11. Reference: NFPA 1031, 4.2.6 and 4.2.6(A)(B)
IFSTA, *Fire Inspection and Code Enforcement,* 6th Edition, 2nd Printing, pages 13–14.
Answer: D

12. Reference: NFPA 1031, 4.2.6 and 4.2.6(A)(B)
IFSTA, *Fire Inspection and Code Enforcement,* 6th Edition, 2nd Printing, pages 13–14.
Answer: C

13. Reference: NFPA 1031, 4.3.1 and 4.3.1(A)(B)
IFSTA, *Fire Inspection and Code Enforcement,* 6th Edition, 1st Printing, page 123.
Answer: D

14. Reference: NFPA 1031, 4.3.1 and 4.3.1(A)(B)
IFSTA, *Fire Inspection and Code Enforcement,* 6th Edition, 1st Printing, page 124.
Answer: A

15. Reference: NFPA 1031, 4.3.10 and 4.3.10(A)(B)
IFSTA, *Fire Inspection and Code Enforcement,* 6th Edition, 1st Printing, page 153.
Answer: D

16. Reference: NFPA 1031, 4.3.10 and 4.3.10(A)(B)
IFSTA, *Fire Inspection and Code Enforcement,* 6th Edition, 1st Printing, page 153.
Answer: D

17. Reference: NFPA 1031, 4.3.11 and 4.3.11(A)(B)
IFSTA, *Fire Inspection and Code Enforcement,* 6th Edition, 1st Printing, page 35.
Answer: A

18. Reference: NFPA 1031, 4.3.11 and 4.3.11(A)(B)
IFSTA, *Fire Inspection and Code Enforcement,* 6th Edition, 1st Printing, pages 136–137.
Answer: B

19. Reference: NFPA 1031, 4.3.12 and 4.3.12(A)(B)
IFSTA, *Fire Inspection and Code Enforcement,* 6th Edition, 1st Printing, page 322.
Answer: A

20. Reference: NFPA 1031, 4.3.12 and 4.3.12(A)(B)
IFSTA, *Fire Inspection and Code Enforcement,* 6th Edition, 1st Printing, page 322.
Answer: D

21. Reference: NFPA 1031, 4.3.12 and 4.3.12(A)(B)
IFSTA, *Fire Inspection and Code Enforcement,* 6th Edition, 1st Printing, page 230.
Answer: A

22. Reference: NFPA 1031, 4.3.12 and 4.3.12(A)(B)
IFSTA, *Fire Inspection and Code Enforcement,* 6th Edition, 1st Printing, page 323.
Answer: C

23. Reference: NFPA 1031, 4.3.12 and 4.3.12(A)(B)
IFSTA, *Fire Inspection and Code Enforcement,* 6th Edition, 1st Printing, pages 323–324.
Answer: C

24. Reference: NFPA 1031, 4.3.12, 4.3.12(A)(B), 4.3.14 and 4.3.14(A)(B)
IFSTA, *Fire Inspection and Code Enforcement,* 6th Edition, 1st Printing, pages 323–324.
Answer: C

25. Reference: NFPA 1031, 4.3.12 and 4.3.12(A)(B)
IFSTA, *Fire Inspection and Code Enforcement,* 6th Edition, 1st Printing, pages 323 and 349.
Answer: C

26. Reference: NFPA 1031, 4.3.12 and 4.3.12(A)(B)
IFSTA, *Fire Inspection and Code Enforcement,* 6th Edition, 1st Printing, page 349.
Answer: B

27. Reference: NFPA 1031, 4.3.12 and 4.3.12(A)(B)
IFSTA, *Fire Inspection and Code Enforcement,* 6th Edition, 1st Printing, page 349.
Answer: B

28. Reference: NFPA 1031, 4.3.12 and 4.3.12(A)(B)
IFSTA, *Fire Inspection and Code Enforcement,* 6th Edition, 1st Printing, page 349.
Answer: A

29. Reference: NFPA 1031, 4.3.12 and 4.3.12(A)(B)
IFSTA, *Fire Inspection and Code Enforcement,* 6th Edition, 1st Printing, page 332.
Answer: D

30. Reference: NFPA 1031, 4.3.12 and 4.3.12(A)(B)
IFSTA, *Fire Inspection and Code Enforcement,* 6th Edition, 1st Printing, page 332.
Answer: A

31. Reference: NFPA 1031, 4.3.12 and 4.3.12(A)(B)
IFSTA, *Fire Inspection and Code Enforcement,* 6th Edition, 1st Printing, page 337.
Answer: A

32. Reference: NFPA 1031, 4.3.12 and 4.3.12(A)(B)
IFSTA, *Fire Inspection and Code Enforcement*, 6th Edition, 1st Printing, page 329.
Answer: C

33. Reference: NFPA 1031, 4.3.13 and 4.3.13(A)(B)
IFSTA, *Fire Inspection and Code Enforcement*, 6th Edition, 1st Printing, page 99.
Answer: D

34. Reference: NFPA 1031, 4.3.13 and 4.3.13(A)(B)
IFSTA, *Fire Inspection and Code Enforcement*, 6th Edition, 1st Printing, page 357.
Answer: B

35. Reference: NFPA 1031, 4.3.14 and 4.3.14(A)(B)
IFSTA, *Essentials of Fire Fighting*, 4th Edition, 1st Printing, page 659.
Answer: D

36. Reference: NFPA 1031, 4.3.14 and 4.3.14(A)(B)
IFSTA, *Fire Inspection and Code Enforcement*, 6th Edition, 1st Printing, page 59.
Answer: B

37. Reference: NFPA 1031, 4.3.14 and 4.3.14(A)(B)
IFSTA, *Fire Inspection and Code Enforcement*, 6th Edition, 1st Printing, page 234.
Answer: B

38. Reference: NFPA 1031, 4.3.14 and 4.3.14(A)(B)
IFSTA, *Fire Inspection and Code Enforcement*, 6th Edition, 1st Printing, page 53.
Answer: C

39. Reference: NFPA 1031, 4.3.14 and 4.3.14(A)(B)
IFSTA, *Essentials of Fire Fighting*, 4th Edition, 1st Printing, page 42.
Answer: B

40. Reference: NFPA 1031, 4.3.14 and 4.3.14(A)(B)
IFSTA, *Fire Inspection and Code Enforcement*, 6th Edition, 1st Printing, page 50.
Answer: D

41. Reference: NFPA 1031, 4.3.14 and 4.3.14(A)(B)
IFSTA, *Fire Inspection and Code Enforcement*, 6th Edition, 1st Printing, page 51.
Answer: B

42. Reference: NFPA 1031, 4.3.14 and 4.3.14(A)(B)
IFSTA, *Fire Inspection and Code Enforcement*, 6th Edition, 1st Printing, page 49.
Answer: C

43. Reference: NFPA 1031, 4.3.14 and 4.3.14(A)(B)
IFSTA, *Fire Inspection and Code Enforcement*, 6th Edition, 1st Printing, page 49.
Answer: A

44. Reference: NFPA 1031, 4.3.14 and 4.3.14(A)(B)
IFSTA, *Fire Inspection and Code Enforcement*, 6th Edition, 1st Printing, page 50.
Answer: A

45. Reference: NFPA 1031, 4.3.14 and 4.3.14(A)(B)
IFSTA, *Fire Inspection and Code Enforcement,* 6th Edition, 1st Printing, page 49.
Answer: A

46. Reference: NFPA 1031, 4.3.14 and 4.3.14(A)(B)
IFSTA, *Fire Inspection and Code Enforcement,* 6th Edition, 1st Printing, page 49.
Answer: B

47. Reference: NFPA 1031, 4.3.14 and 4.3.14(A)(B)
IFSTA, *Fire Inspection and Code Enforcement,* 6th Edition, 1st Printing, pages 55–56.
Answer: D

48. Reference: NFPA 1031, 4.3.14 and 4.3.14(A)(B)
IFSTA, *Fire Inspection and Code Enforcement,* 6th Edition, 1st Printing, pages 55–56.
Answer: A

49. Reference: NFPA 1031, 4.3.14 and 4.3.14(A)(B)
IFSTA, *Fire Inspection and Code Enforcement,* 6th Edition, 1st Printing, page 49.
Answer: B

50. Reference: NFPA 1031, 4.3.16 and 4.3.16(A)(B)
IFSTA, *Fire Inspection and Code Enforcement,* 6th Edition, 2nd Printing, page 209.
Answer: D

51. Reference: NFPA 1031, 4.3.16 and 4.3.16(A)(B)
IFSTA, *Fire Inspection and Code Enforcement,* 6th Edition, 2nd Printing, page 204.
Answer: D

52. Reference: NFPA 1031, 4.3.2, 4.3.2(A)(B), 4.3.1, and 4.3.1(A)(B)
IFSTA, *Fire Inspection and Code Enforcement,* 6th Edition, 1st Printing, page 146.
Answer: B

53. Reference: NFPA 1031, 4.3.2, 4.3.2(A)(B), 4.3.1, and 4.3.1(A)(B)
IFSTA, *Fire Inspection and Code Enforcement,* 6th Edition, 1st Printing, page 146.
Answer: A

54. Reference: NFPA 1031, 4.3.3, 4.3.3(A)(B), 4.2.1, and 4.2.1(A)(B)
IFSTA, *Fire Inspection and Code Enforcement,* 6th Edition, 1st Printing, page 31.
Answer: D

55. Reference: NFPA 1031, 4.3.3, 4.3.3(A)(B), 4.2.5, and 4.2.5(A)(B)
IFSTA, *Fire Inspection and Code Enforcement,* 6th Edition, 1st Printing, page 66.
Answer: C

56. Reference: NFPA 1031, 4.3.3, 4.3.3(A)(B), 4.3.15, 4.3.15(A)(B), 4.3.14, and 4.3.14(A)(B)
IFSTA, *Fire Inspection and Code Enforcement,* 6th Edition, 1st Printing, page 140.
Answer: D

57. Reference: NFPA 1031, 4.3.3, 4.3.3(A)(B), 4.2.5, and 4.2.5(A)(B)
IFSTA, *Fire Inspection and Code Enforcement,* 6th Edition, 1st Printing, page 139.
Answer: D

58. Reference: NFPA 1031, 4.3.3, 4.3.3(A)(B), 4.2.5, and 4.2.5(A)(B)
IFSTA, *Fire Inspection and Code Enforcement,* 6th Edition, 1st Printing, page 150, Table 6.3.
Answer: B

59. Reference: NFPA 1031, 4.3.5 and 4.3.5(A)
IFSTA, *Fire Inspection and Code Enforcement,* 6th Edition, 1st Printing, pages 252–254.
Answer: B

60. Reference: NFPA 1031, 4.3.5, 4.3.5(A)(B), 4.2.5, 4.2.5(A)(B), 4.3.4, and 4.3.4(A)(B)
IFSTA, *Fire Inspection and Code Enforcement,* 6th Edition, 1st Printing, page 180.
Answer: E

61. Reference: NFPA 1031, 4.3.5, 4.3.5(A)(B), 4.2.5, 4.2.5(A)(B), 4.3.4, and 4.3.4(A)(B)
IFSTA, *Fire Inspection and Code Enforcement,* 6th Edition, 1st Printing, page 180.
Answer: B

62. Reference: NFPA 1031, 4.3.5, 4.3.5(A)(B), 4.3.16, and 4.3.16(A)(B)
IFSTA, *Fire Inspection and Code Enforcement,* 6th Edition, 1st Printing, page 208.
Answer: B

63. Reference: NFPA 1031, 4.3.5, 4.3.5(A)(B), 4.3.16, and 4.3.16(A)(B)
IFSTA, *Fire Inspection and Code Enforcement,* 6th Edition, 1st Printing, page 453.
Answer: B

64. Reference: NFPA 1031, 4.3.5, 4.3.5(A)(B), 4.3.16, and 4.3.16(A)(B)
IFSTA, *Fire Inspection and Code Enforcement,* 6th Edition, 1st Printing, page 452.
Answer: A

65. Reference: NFPA 1031, 4.3.5, 4.3.5(A), 4.3.16, and 4.3.16(A)
IFSTA, *Essentials of Fire Fighting,* 4th Edition, 1st Printing, page 490.
Answer: C

66. Reference: NFPA 1031, 4.3.5, 4.3.5(A)(B), 4.3.16, and 4.3.16(A)(B)
IFSTA, *Fire Inspection and Code Enforcement,* 6th Edition, 1st Printing, page 160.
Answer: A

67. Reference: NFPA 1031, 4.3.5, 4.3.5(A)(B), 4.3.16, and 4.3.16(A)(B)
IFSTA, *Fire Inspection and Code Enforcement,* 6th Edition, 1st Printing, pages 160–161.
Answer: A

68. Reference: NFPA 1031, 4.3.5, 4.3.5(A)(B), 4.3.16, and 4.3.16(A)(B)
IFSTA, *Fire Inspection and Code Enforcement,* 6th Edition, 1st Printing, page 163.
Answer: D

69. Reference: NFPA 1031, 4.3.5, 4.3.5(A), 4.3.16, and 4.3.16(A)(B)
IFSTA, *Fire Inspection and Code Enforcement,* 6th Edition, 1st Printing, page 159.
Answer: D

70. Reference: NFPA 1031, 4.3.5, 4.3.5(A)(B), 4.3.16, and 4.3.16(A)(B)
IFSTA, *Fire Inspection and Code Enforcement,* 6th Edition, 1st Printing, page 174.
Answer: B

71. Reference: NFPA 1031, 4.3.5 and 4.3.5(A)(B), 4.3.16, and 4.3.16(A)(B)
IFSTA, *Fire Inspection and Code Enforcement,* 6th Edition, 1st Printing, pages 162–163.
Answer: D

72. Reference: NFPA 1031, 4.3.5, 4.3.5(A)(B), 4.3.16, and 4.3.16(A)(B)
IFSTA, *Fire Inspection and Code Enforcement,* 6th Edition, 1st Printing, page 173.
Answer: A

73. Reference: NFPA 1031, 4.3.5, 4.3.5(A)(B), 4.3.16, and 4.3.16(A)(B)
IFSTA, *Fire Inspection and Code Enforcement,* 6th Edition, 1st Printing, pages 162–163.
Answer: C

74. Reference: NFPA 1031, 4.3.5, 4.3.5(A)(B), 4.3.16, and 4.3.16(A)(B)
IFSTA, *Fire Inspection and Code Enforcement,* 6th Edition, 1st Printing, page 163.
Answer: D

75. Reference: NFPA 1031, 4.3.5, 4.3.5(A)(B), 4.3.16, and 4.3.16(A)(B)
IFSTA, *Fire Inspection and Code Enforcement,* 6th Edition, 1st Printing, page 163.
Answer: B

76. Reference: NFPA 1031, 4.3.5, 4.3.5(A)(B), 4.3.16, and 4.3.16((A)(B)
IFSTA, *Fire Inspection and Code Enforcement,* 6th Edition, 1st Printing, page 163.
Answer: B

77. Reference: NFPA 1031, 4.3.5, 4.3.5(A)(B), 4.3.16, and 4.3.16(A)(B)
IFSTA, *Fire Inspection and Code Enforcement,* 6th Edition, 1st Printing, page 163.
Answer: B

78. Reference: NFPA 1031, 4.3.5, 4.3.5(A)(B), 4.3.16, and 4.3.16(A)(B)
IFSTA, *Fire Inspection and Code Enforcement,* 6th Edition, 1st Printing, pages 160–161, p. 160 Figure 7.3.
Answer: D

79. Reference: NFPA 1031, 4.3.5, 4.3.5(A)(B), 4.3.16, and 4.3.16(A)(B)
IFSTA, *Fire Inspection and Code Enforcement,* 6th Edition, 1st Printing, page 164.
Answer: D

80. Reference: NFPA 1031, 4.3.6 and 4.3.6(A)(B)
IFSTA, *Fire Inspection and Code Enforcement,* 6th Edition, 1st Printing, pages 306–307.
Answer: D

81. Reference: NFPA 1031, 4.3.6 and 4.3.6(A)(B)
IFSTA, *Fire Inspection and Code Enforcement,* 6th Edition, 1st Printing, page 280.
Answer: C

82. Reference: NFPA 1031, 4.3.7 and 4.3.7(A)
IFSTA, *Fire Inspection and Code Enforcement,* 6th Edition, 1st Printing, page 230.
Answer: C

83. Reference: NFPA 1031, 4.3.7 and 4.3.7(A)
IFSTA, *Fire Inspection and Code Enforcement,* 6th Edition, 1st Printing, page 234, Figure 8-11.
Answer: C

84. Reference: NFPA 1031, 4.3.7 and 4.3.7(A)
IFSTA, *Fire Inspection and Code Enforcement,* 6th Edition, 1st Printing, page 234, Figure 8-11.
Answer: B

85. Reference: NFPA 1031, 4.3.7 and 4.3.7(A)(B)
IFSTA, *Fire Inspection and Code Enforcement,* 6th Edition, 1st Printing, page 232.
Answer: B

86. Reference: NFPA 1031, 4.3.7 and 4.3.7(A)(B)
IFSTA, *Fire Inspection and Code Enforcement,* 6th Edition, 1st Printing, pages 230 and 238.
Answer: B

87. Reference: NFPA 1031, 4.3.7 and 4.3.7(A)(B)
IFSTA, *Fire Inspection and Code Enforcement,* 6th Edition, 1st Printing, page 234.
Answer: C

88. Reference: NFPA 1031, 4.3.7 and 4.3.7(A)(B)
IFSTA, *Fire Inspection and Code Enforcement,* 6th Edition, 1st Printing, page 244.
Answer: B

89. Reference: NFPA 1031, 4.3.7 and 4.3.7(A)(B)
IFSTA, *Fire Inspection and Code Enforcement,* 6th Edition, 1st Printing, pages 231–323.
Answer: A

90. Reference: NFPA 1031, 4.3.7 and 4.3.7(A)(B)
IFSTA, *Fire Inspection and Code Enforcement,* 6th Edition, 1st Printing, page 242.
Answer: D

91. Reference: NFPA 1031, 4.3.7 and 4.3.7(A)(B)
IFSTA, *Fire Inspection and Code Enforcement,* 6th Edition, 1st Printing, page 234.
Answer: B

92. Reference: NFPA 1031, 4.3.7 and 4.3.7(A)(B)
IFSTA, *Fire Inspection and Code Enforcement,* 6th Edition, 1st Printing, page 237.
Answer: D

93. Reference: NFPA 1031, 4.3.8 and 4.3.8(A)
IFSTA, *Fire Inspection and Code Enforcement,* 6th Edition, 1st Printing, page 93.
Answer: D

94. Reference: NFPA 1031, 4.3.8 and 4.3.8(A)(B)
IFSTA, *Fire Inspection and Code Enforcement,* 6th Edition, 1st Printing, page 82.
Answer: A

95. Reference: NFPA 1031, 4.3.8 and 4.3.8(A)(B)
IFSTA, *Fire Inspection and Code Enforcement,* 6th Edition, 1st Printing, page 453.
Answer: A

96. Reference: NFPA 1031, 4.3.8 and 4.3.8(A)(B)
IFSTA, *Fire Inspection and Code Enforcement,* 6th Edition, 1st Printing, pages 75–76, Table 4.1.

Answer: A

97. Reference: NFPA 1031, 4.3.8 and 4.3.8(A)(B)
IFSTA, *Fire Inspection and Code Enforcement,* 6th Edition, 1st Printing, pages 77–78.
Answer: B

98. Reference: NFPA 1031, 4.3.9, 4.3.9(A)(B), 4.3.8, and 4.3.8(A)(B)
IFSTA, *Fire Inspection and Code Enforcement,* 6th Edition, 1st Printing, pages 69 and 90.
Answer: D

99. Reference: NFPA 1031, 4.3.9, 4.3.9(A)(B), 4.3.10, 4.3.10(A)(B), 4.3.14, and 4.3.14(A)(B)
IFSTA, *Fire Inspection and Code Enforcement,* 6th Edition, 1st Printing, page 289.
Answer: D

100. Reference: NFPA 1031, 4.3.9, 4.3.9(A)(B), 4.3.10, and 4.3.10(A)(B)
IFSTA, *Fire Inspection and Code Enforcement,* 6th Edition, 1st Printing, pages 292–293.
Answer: B

Examination I-2 Answer Key

Directions

Follow these steps carefully for completing the feedback part of the Systematic Approach to Examination Preparation (SAEP):

1. After entering your scores, look up the answers for the examination items you missed as well as those you guessed, even if you guessed correctly. If you are guessing, it means the answer isn't perfectly clear. This process will make you as knowledgeable as possible.
2. Enter the number of missed and guessed examination items in the blank on your Personal Progress Plotter.
3. Highlight the answer in the reference materials, and then read the paragraph preceding and the paragraph following the one in which the correct answer is located. Enter the paragraph number and page number next to the guessed or missed examination item on your examination. Count any part of a paragraph at the beginning of the page as one paragraph until you reach the paragraph containing your highlighted answer. This step will help you locate and review your missed and guessed examination items later in the process. This step is essential to learning the material in context and by association. These learning techniques (context/association) are the very backbone of the SAEP approach.
4. Once you have completed the feedback part, you may proceed to the next examination.

1. Reference: NFPA 1031, 4.2.1, 4.2.1(A)(B), 4.2.6, and 4.2.6(A)(B)
   IFSTA, *Fire Inspection and Code Enforcement,* 6th Edition, 2nd Printing, page 36.
   Answer: B

2. Reference: NFPA 1031, 4.2.1 and 4.2.1(A)(B)
   IFSTA, *Fire Inspection and Code Enforcement,* 6th Edition, 1st Printing, page 36.
   Answer: B

3. Reference: NFPA 1031, 4.2.1 and 4.2.1(A)(B)
   IFSTA, *Fire Inspection and Code Enforcement,* 6th Edition, 1st Printing, page 36.
   Answer: C

4. Reference: NFPA 1031, 4.2.1, 4.2.1(A)(B), 4.3.1, and 4.3.1(A)(B)
   IFSTA, *Fire Inspection and Code Enforcement,* 6th Edition, 1st Printing, page 36.
   Answer: C

5. Reference: NFPA 1031, 4.2.2 and 4.2.2(A)(B)
   IFSTA, *Fire Inspection and Code Enforcement,* 6th Edition, 2nd Printing, pages 19–20.
   Answer: C

6. Reference: NFPA 1031, 4.2.3, 4.2.3(A), 4.3.9, and 4.3.9(A)(B)
   IFSTA, *Fire Inspection and Code Enforcement,* 6th Edition, 2nd Printing, pages 290–292.
   Answer: D

7. Reference: NFPA 1031, 4.2.3, 4.2.3(A), 4.3.9, and 4.3.9(A)(B)
IFSTA, *Fire Inspection and Code Enforcement,* 6th Edition, 2nd Printing, pages 289 and 452.
Answer: C

8. Reference: NFPA 1031, 4.2.4 and 4.2.4(A)(B)
IFSTA, *Fire Inspection and Code Enforcement,* 6th Edition, 2nd Printing, pages 14–15.
Answer: B

9. Reference: NFPA 1031, 4.2.5 and 4.2.5(A)(B)
IFSTA, *Fire Inspection and Code Enforcement,* 6th Edition, 2nd Printing, pages 16–17.
Answer: B

10. Reference: NFPA 1031, 4.2.6 and 4.2.6(A)(B)
IFSTA, *Fire Inspection and Code Enforcement,* 6th Edition, 2nd Printing, page 13.
Answer: C

11. Reference: NFPA 1031, 4.2.6 and 4.2.6(A)(B)
IFSTA, *Fire Inspection and Code Enforcement,* 6th Edition, 2nd Printing, page 13.
Answer: A

12. Reference: NFPA 1031, 4.2.6 and 4.2.6(A)(B)
IFSTA, *Fire Inspection and Code Enforcement,* 6th Edition, 2nd Printing, page 13.
Answer: B

13. Reference: NFPA 1031, 4.3.1 and 4.3.1(A)(B)
IFSTA, *Fire Inspection and Code Enforcement,* 6th Edition, 2nd Printing, page 123.
Answer: B

14. Reference: NFPA 1031, 4.3.1 and 4.3.1(A)(B)
IFSTA, *Fire Inspection and Code Enforcement,* 6th Edition, 1st Printing, page 122.
Answer: A

15. Reference: NFPA 1031, 4.3.10 and 4.3.10(A)(B)
IFSTA, *Fire Inspection and Code Enforcement,* 6th Edition, 1st Printing, page 153.
Answer: B

16. Reference: NFPA 1031, 4.3.10 and 4.3.10(A)(B)
IFSTA, *Fire Inspection and Code Enforcement,* 6th Edition, 2nd Printing, page 153.
Answer: D

17. Reference 1031, NFPA 4.3.11 and 4.3.11(A)(B)
IFSTA, *Fire Inspection and Code Enforcement,* 6th Edition, 1st Printing, page 136.
Answer: A

18. Reference: NFPA 1031, 4.3.12 and 4.3.12(A)(B)
IFSTA, *Fire Inspection and Code Enforcement,* 6th Edition, 1st Printing, page 337.
Answer: C

19. Reference: NFPA 1031, 4.3.12 and 4.3.12(A)(B)
IFSTA, *Fire Inspection and Code Enforcement,* 6th Edition, 1st Printing, pages 325–326.
Answer: D

20. Reference: NFPA 1031, 4.3.1.2 and 4.3.12(A)(B)
IFSTA, *Fire Inspection and Code Enforcement,* 6th Edition, 1st Printing, page 332.
Answer: D

21. Reference: NFPA 1031, 4.3.12 and 4.3.12(A)(B)
IFSTA, *Fire Inspection and Code Enforcement,* 6th Edition, 1st Printing, page 328.
Answer: C

22. Reference: NFPA 1031, 4.3.12 and 4.3.12(A)(B)
IFSTA, *Fire Inspection and Code Enforcement,* 6th Edition, 1st Printing, page 331.
Answer: B

23. Reference: NFPA 1031, 4.3.12 and 4.3.12(A)(B)
IFSTA, *Fire Inspection and Code Enforcement,* 6th Edition, 1st Printing, page 331.
Answer: A

24. Reference: NFPA 1031, 4.3.12 and 4.3.12(A)(B)
IFSTA, *Fire Inspection and Code Enforcement,* 6th Edition, 1st Printing, page 338.
Answer: B

25. Reference: NFPA 1031, 4.3.12 and 4.3.12(A)(B)
IFSTA, *Fire Inspection and Code Enforcement,* 6th Edition, 1st Printing, page 338.
Answer: A

26. Reference: NFPA 1031, 4.3.12 and 4.3.12(A)(B)
IFSTA, *Fire Inspection and Code Enforcement,* 6th Edition, 1st Printing, page 325.
Answer: C

27. Reference: NFPA 1031, 4.3.12 and 4.3.12(A)(B)
IFSTA, *Fire Inspection and Code Enforcement,* 6th Edition, 1st Printing, page 331.
Answer: C

28. Reference: NFPA 1031, 4.3.12 and 4.3.12(A)(B)
IFSTA, *Fire Inspection and Code Enforcement,* 6th Edition, 1st Printing, page 327.
Answer: C

29. Reference: NFPA 1031, 4.3.12 and 4.3.12(A)(B)
IFSTA, *Fire Inspection and Code Enforcement,* 6th Edition, 1st Printing, page 331.
Answer: C

30. Reference: NFPA 1031, 4.3.12 and 4.3.12(A)(B)
IFSTA, *Fire Inspection and Code Enforcement,* 6th Edition, 1st Printing, page 336.
Answer: A

31. Reference: NFPA 1031, 4.3.12 and 4.3.12(A)(B)
IFSTA, *Fire Inspection and Code Enforcement,* 6th Edition, 1st Printing, page 334.
Answer: D

32. Reference: NFPA 1031, 4.3.12 and 4.3.12(A)(B)
IFSTA, *Fire Inspection and Code Enforcement,* 6th Edition, 1st Printing, pages 338–339.

Answer: C

33. Reference: NFPA 1031, 4.3.13 and 4.3.13(A)(B)
IFSTA, *Fire Inspection and Code Enforcement,* 6th Edition, 1st Printing, page 356.
Answer: C

34. Reference: NFPA 1031, 4.3.13 and 4.3.13(A)(B)
IFSTA, *Fire Inspection and Code Enforcement,* 6th Edition, 1st Printing, pages 100–101.
Answer: C

35. Reference: NFPA 1031, 4.3.13 and 4.3.13(A)(B)
IFSTA, *Fire Inspection and Code Enforcement,* 6th Edition, 1st Printing, page 102.
Answer: A

36. Reference: NFPA 1031, 4.3.14 and 4.3.14(A)(B)
IFSTA, *Fire Inspection and Code Enforcement,* 6th Edition, 1st Printing, pages 53 and 56.
Answer: B

37. Reference: NFPA 1031, 4.3.14 and 4.3.14(A)(B)
IFSTA, *Fire Inspection and Code Enforcement,* 6th Edition, 1st Printing, page 56.
Answer: A

38. Reference: NFPA 1031, 4.3.14 and 4.3.14(A)(B)
IFSTA, *Fire Inspection and Code Enforcement,* 6th Edition, 1st Printing, page 50.
Answer: B

39. Reference: NFPA 1031, 4.3.14 and 4.3.14(A)(B)
IFSTA, *Fire Inspection and Code Enforcement,* 6th Edition, 1st Printing, page 54.
Answer: A

40. Reference: NFPA 1031, 4.3.14 and 4.3.14(A)(B)
IFSTA, *Fire Inspection and Code Enforcement,* 6th Edition, 1st Printing, page 54.
Answer: D

41. Reference: NFPA 1031, 4.3.14 and 4.3.14(A)(B)
IFSTA, *Fire Inspection and Code Enforcement,* 6th Edition, 1st Printing, page 48.
Answer: A

42. Reference: NFPA 1031, 4.3.14 and 4.3.14(A)(B)
IFSTA, *Fire Inspection and Code Enforcement,* 6th Edition, 1st Printing, page 54.
Answer: C

43. Reference: NFPA 1031, 4.3.14 and 4.3.14(A)(B)
IFSTA, *Fire Inspection and Code Enforcement,* 6th Edition, 1st Printing, page 56.
Answer: C

44. Reference: NFPA 1031, 4.3.14 and 4.3.14(A)(B)
IFSTA, *Fire Inspection and Code Enforcement,* 6th Edition, 1st Printing, page 54.
Answer: C

45. Reference: NFPA 1031, 4.3.14 and 4.3.14(A)(B)
IFSTA, *Fire Inspection and Code Enforcement,* 6th Edition, 1st Printing, page 53.
Answer: B

46. Reference: NFPA 1031, 4.3.14 and 4.3.14(A)(B)
IFSTA, *Fire Inspection and Code Enforcement,* 6th Edition, 1st Printing, page 48.
Answer: C

47. Reference: NFPA 1031, 4.3.14 and 4.3.14(A)(B)
IFSTA, *Fire Inspection and Code Enforcement,* 6th Edition, 1st Printing, page 48.
Answer: B

48. Reference: NFPA 1031, 4.3.14 and 4.3.14(A)(B)
IFSTA, *Fire Inspection and Code Enforcement,* 6th Edition, 1st Printing, page 47.
Answer: B

49. Reference: NFPA 1031, 4.3.14 and 4.3.14(A)(B)
IFSTA, *Fire Inspection and Code Enforcement,* 6th Edition, 1st Printing, page 56.
Answer: B

50. Reference: NFPA 1031, 4.3.14 and 4.3.14(A)(B)
IFSTA, *Fire Inspection and Code Enforcement,* 6th Edition, 1st Printing, pages 56 and 448.
Answer: D

51. Reference: NFPA 1031, 4.3.16 and 4.3.16(A)(B)
IFSTA, *Fire Inspection and Code Enforcement,* 6th Edition, 2nd Printing, page 208.
Answer: C

52. Reference: NFPA 1031, 4.3.16 and 4.3.16(A)
IFSTA, *Fire Inspection and Code Enforcement,* 6th Edition, 2nd Printing, page 208.
Answer: B

53. Reference: NFPA 1031, 4.3.2 and 4.3.2(A)
IFSTA, *Fire Inspection and Code Enforcement,* 6th Edition, 2nd Printing, page 145.
Answer: B

54. Reference: NFPA 1031, 4.3.2, 4.3.2(A), 4.3.1, and 4.3.1(A)(B)
IFSTA, *Fire Inspection and Code Enforcement,* 6th Edition, 2nd Printing, pages 146–147.
Answer: B

55. Reference: NFPA 1031, 4.3.3, 4.3.3(A)(B), 4.2.5, and 4.2.5(A)(B)
IFSTA, *Fire Inspection and Code Enforcement,* 6th Edition, 2nd Printing, page 138.
Answer: B

56. Reference: NFPA 1031, 4.3.3 and 4.3.3(A)
IFSTA, *Fire Inspection and Code Enforcement,* 6th Edition, 2nd Printing, page 143.
Answer: B

57. Reference: NFPA 1031, 4.3.4 and 4.3.4(A)(B)
IFSTA, *Fire Inspection and Code Enforcement,* 6th Edition, 2nd Printing, page 117.
Answer: C

58. Reference: NFPA 1031, 4.3.4 and 4.3.4(A)(B)
IFSTA, *Fire Inspection and Code Enforcement,* 6th Edition, 1st Printing, page 69.
Answer: D

59. Reference: NFPA 1031, 4.3.4 and 4.3.4(A)(B)
IFSTA, *Fire Inspection and Code Enforcement,* 6th Edition, 1st Printing, page 64.
Answer: B

60. Reference: NFPA 1031, 4.3.4 and 4.3.4(A)(B)
IFSTA, *Fire Inspection and Code Enforcement,* 6th Edition, 1st Printing, pages 69–70.
Answer: A

61. Reference: NFPA 1031, 4.3.5, 4.3.5(A)(B), 4.3.16, and 4.3.16(A)(B)
IFSTA, *Fire Inspection and Code Enforcement,* 6th Edition, 1st Printing, page 163.
Answer: C

62. Reference: NFPA 1031, 4.3.5, 4.3.5(A)(B), 4.3.16, and 4.3.16(A)(B)
IFSTA, *Fire Inspection and Code Enforcement,* 6th Edition, 1st Printing, page 206.
Answer: B

63. Reference: NFPA 1031, 4.3.5, 4.3.5(A)(B), 4.3.16, and 4.3.16(A)(B)
IFSTA, *Fire Inspection and Code Enforcement,* 6th Edition, 1st Printing, page 208.
Answer: A

64. Reference: NFPA 1031, 4.3.5, 4.3.5(A)(B), 4.3.16, and 4.3.16(A)(B)
IFSTA, *Fire Inspection and Code Enforcement,* 6th Edition, 1st Printing, page 205.
Answer: C

65. Reference: NFPA 1031, 4.3.5, 4.3.5(A)(B), 4.3.16, and 4.3.16(A)(B)
IFSTA, *Fire Inspection and Code Enforcement,* 6th Edition, 1st Printing, page 209.
Answer: B

66. Reference: NFPA 1031, 4.3.5, 4.3.5(A)(B), 4.3.16, and 4.3.16(A)(B)
IFSTA, *Fire Inspection and Code Enforcement,* 6th Edition, 1st Printing, pages 206–207.
Answer: D

67. Reference: NFPA 1031, 4.3.5, 4.3.5(A), 4.3.16, and 4.3.16(A)(B)
IFSTA, *Fire Inspection and Code Enforcement,* 6th Edition, 1st Printing, page 206.
Answer: D

68. Reference: NFPA 1031, 4.3.5, 4.3.5(A)(B), 4.3.16, and 4.3.16(A)(B)
IFSTA, *Fire Inspection and Code Enforcement,* 6th Edition, 1st Printing, page 217.
Answer: D

69. Reference: NFPA 1031, 4.3.5, 4.3.5(A)(B), 4.3.16, and 4.3.16(A)(B)
IFSTA, *Fire Inspection and Code Enforcement,* 6th Edition, 1st Printing, page 453.
Answer: B

70. Reference: NFPA 1031, 4.3.5, 4.3.5(A)(B), 4.3.16, and 4.3.16(A)(B)
IFSTA, *Fire Inspection and Code Enforcement,* 6th Edition, 1st Printing, pages 214–216.
Answer: A

71. Reference: NFPA 1031, 4.3.5, 4.3.5(A)(B), 4.3.16, and 4.3.16(A)(B)
IFSTA, *Fire Inspection and Code Enforcement,* 6th Edition, 1st Printing, page 208.
Answer: C

72. Reference: NFPA 1031, 4.3.5, 4.3.5(A)(B), 4.3.16, and 4.3.16(A)(B)
IFSTA, *Fire Inspection and Code Enforcement,* 6th Edition, 1st Printing, page 208.
Answer: C

73. Reference: NFPA 1031, 4.3.5, 4.3.5(A)(B), 4.3.16, and 4.3.16(A)(B)
IFSTA, *Fire Inspection and Code Enforcement,* 6th Edition, 1st Printing, pages 206–207.
Answer: D

74. Reference: NFPA 1031, 4.3.5, 4.3.5(A)(B), 4.3.16, and 4.3.16(A)(B)
IFSTA, *Fire Inspection and Code Enforcement,* 6th Edition, 1st Printing, page 213.
Answer: D

75. Reference: NFPA 1031, 4.3.5, 4.3.5(A)(B), 4.3.16, and 4.3.16(A)(B)
IFSTA, *Fire Inspection and Code Enforcement,* 6th Edition, 1st Printing, page 453.
IFSTA, *Essentials of Fire Fighting,* 4th Edition, 1st Printing, pages 385–386.
Answer: A

76. Reference: NFPA 1031, 4.3.5, 4.3.5(A), 4.3.16, and 4.3.16(A)(B)
IFSTA, *Fire Inspection and Code Enforcement,* 6th Edition, 1st Printing, page 190.
Answer: C

77. Reference: NFPA 1031, 4.3.5, 4.3.5(A)(B), 4.3.16, and 4.3.16(A)(B)
IFSTA, *Fire Inspection and Code Enforcement,* 6th Edition, 1st Printing, page 171.
Answer: B

78. Reference: NFPA 1031, 4.3.5, 4.3.5(A)(B), 4.3.16, and 4.3.16(A)(B)
IFSTA, *Fire Inspection and Code Enforcement,* 6th Edition, 1st Printing, page 192.
Answer: D

79. Reference: NFPA 1031, 4.3.5, 4.3.5(A)(B), 4.3.16, and 4.3.16(A)(B)
IFSTA, *Fire Inspection and Code Enforcement,* 6th Edition, 1st Printing, pages 162–163.
Answer: C

80. Reference: NFPA 1031, 4.3.5, 4.3.5(A)(B), 4.3.16, and 4.3.16(A)(B)
IFSTA, *Fire Inspection and Code Enforcement,* 6th Edition, 1st Printing, page 162.
Answer: C

81. Reference: NFPA 1031, 4.3.5, and 4.3.5(A)(B)
IFSTA, *Fire Inspection and Code Enforcement,* 6th Edition, 1st Printing, pages 187–188.
Answer: C

82. Reference: NFPA 1031, 4.3.6 and 4.3.6(A)(B)
IFSTA, *Fire Inspection and Code Enforcement,* 6th Edition, 1st Printing, page 278.
Answer: A

83. Reference: NFPA 1031, 4.3.6 and 4.3.6(A)(B)
IFSTA, *Fire Inspection and Code Enforcement,* 6th Edition, 1st Printing, page 276.
Answer: D

84. Reference: NFPA 1031, 4.3.6 and 4.3.6(A)(B)
IFSTA, *Fire Inspection and Code Enforcement,* 6th Edition, 1st Printing, page 270.
Answer: D

85. Reference: NFPA 1031, 4.3.7 and 4.3.7(A)(B)
IFSTA, *Fire Inspection and Code Enforcement,* 6th Edition, 1st Printing, pages 238–239.
Answer: C

86. Reference: NFPA 1031, 4.3.7 and 4.3.7(A)(B)
IFSTA, *Fire Inspection and Code Enforcement,* 6th Edition, 1st Printing, page 234, Figure 8.11.
Answer: C

87. Reference: NFPA 1031, 4.3.7 and 4.3.7(A)(B)
IFSTA, *Fire Inspection and Code Enforcement,* 6th Edition, 1st Printing, page 234, Figure 8.11.
Answer: D

88. Reference: NFPA 1031, 4.3.7 and 4.3.7(A)(B)
IFSTA, *Fire Inspection and Code Enforcement,* 6th Edition, 1st Printing, page 237.
Answer: D

89. Reference: NFPA 1031, 4.3.7 and 4.3.7(A)(B)
IFSTA, *Fire Inspection and Code Enforcement,* 6th Edition, 1st Printing, pages 231, 233, and 234.
Answer: A

90. Reference: NFPA 1031, 4.3.7 and 4.3.7(A)(B)
IFSTA, *Fire Inspection and Code Enforcement,* 6th Edition, 1st Printing, pages 230, 232, and 234.
Answer: A

91. Reference: NFPA 1031, 4.3.7 and 4.3.7(A)(B)
IFSTA, *Fire Inspection and Code Enforcement,* 6th Edition, 1st Printing, page 234, Figure 8.11.
Answer: B

92. Reference: NFPA 1031, 4.3.7 and 4.3.7(A)(B)
IFSTA, *Fire Inspection and Code Enforcement,* 6th Edition, 1st Printing, pages 230 and 234.
Answer: A

93. Reference: NFPA 1031, 4.3.7 and 4.3.7(A)(B)
IFSTA, *Fire Inspection and Code Enforcement,* 6th Edition, 1st Printing, pages 231 and 234.
Answer: D

94. Reference: NFPA 1031, 4.3.7 and 4.3.7(A)(B)
IFSTA, *Fire Inspection and Code Enforcement,* 6th Edition, 1st Printing, page 237.
Answer: D

95. Reference: NFPA 1031, 4.3.7 and 4.3.7(A)(B)
IFSTA, *Fire Inspection and Code Enforcement,* 6th Edition, 1st Printing, page 229.
Answer: C

96. Reference: NFPA 1031, 4.3.8 and 4.3.8(A)
IFSTA, *Fire Inspection and Code Enforcement,* 6th Edition, 1st Printing, pages 93–94.
Answer: C

97. Reference: NFPA 1031, 4.3.8, 4.3.8(A)(B), 4.2.5, and 4.2.5(A)(B)
IFSTA, *Fire Inspection and Code Enforcement,* 6th Edition, 1st Printing, page 97.
Answer: B

98. Reference: NFPA 1031, 4.3.8, 4.3.8(A)(B), 4.2.5, and 4.2.5(A)(B)
IFSTA, *Fire Inspection and Code Enforcement,* 6th Edition, 1st Printing, page 94.
Answer: B

99. Reference: NFPA 1031, 4.3.9, 4.3.9(A)(B), 4.3.8, and 4.3.8(A)(B)
IFSTA, *Fire Inspection and Code Enforcement,* 6th Edition, 1st Printing, page 67.
Answer: A

100. Reference: NFPA 1031, 4.3.9, 4.3.9(A)(B), 4.3.15, and 4.3.15(A)(B)
IFSTA, *Fire Inspection and Code Enforcement,* 6th Edition, 1st Printing, page 289.
Answer: B

Examination I-3 Answer Key

Directions

Follow these steps carefully for completing the feedback part of the Systematic Approach to Examination Preparation (SAEP):

1. After entering your scores, look up the answers for the examination items you missed as well as those you guessed, even if you guessed correctly. If you are guessing, it means the answer isn't perfectly clear. This process will make you as knowledgeable as possible.
2. Enter the number of missed and guessed examination items in the blank on your Personal Progress Plotter.
3. Highlight the answer in the reference materials, and then read the paragraph preceding and the paragraph following the one in which the correct answer is located. Enter the paragraph number and page number next to the guessed or missed examination item on your examination. Count any part of a paragraph at the beginning of the page as one paragraph until you reach the paragraph containing your highlighted answer. This step will help you locate and review your missed and guessed examination items later in the process. This step is essential to learning the material in context and by association. These learning techniques (context/association) are the very backbone of the SAEP approach.
4. **Congratulations!** You have completed the examination and feedback parts of SAEP when you have highlighted your guessed and missed examination items for this examination.

Proceed to Phase III and Phase IV. Study the materials carefully in these important phases. They will help you polish your examination-taking skills. Approximately two to three days prior to taking your next examination, carefully read all of the highlighted information in the reference materials, using the same techniques applied during the feedback part. This exercise will reinforce your learning and provide you with an added level of confidence going into the examination.

Someone once said to professional golfer Tom Watson after he won several tournament championships, "You are really lucky to have won those championships. You are really on a streak." Watson was reported to have replied, "Yes, there is some luck involved, but what I've really noticed is that the more I practice, the luckier I get." What Watson was saying is that good luck usually results from good preparation. This line of thinking certainly applies to learning the rules and hints of examination taking.

Rule 7

Good luck = Good preparation.

1. Reference: NFPA 1031, 4.2.1 and 4.2.1(A)(B)
   IFSTA, *Fire Inspection and Code Enforcement*, 6th Edition, 1st Printing, page 13.
   Answer: C

2. Reference: NFPA 1031, 4.2.1 and 4.2.1(A)(B)
   IFSTA, *Fire Inspection and Code Enforcement*, 6th Edition, 1st Printing, page 36.
   Answer: B

3. Reference: NFPA 1031, 4.2.1 and 4.2.1(A)(B)
IFSTA, *Fire Inspection and Code Enforcement,* 6th Edition, 1st Printing, page 36.
Answer: C

4. Reference: NFPA 1031, 4.2.1 and 4.2.1(A)(B)
IFSTA, *Fire Inspection and Code Enforcement,* 6th Edition, 1st Printing, page 30.
Answer: A

5. Reference: NFPA 1031, 4.2.1 and 4.2.1(A)(B)
IFSTA, *Fire Inspection and Code Enforcement,* 6th Edition, 2nd Printing, page 42.
Answer: D

6. Reference: NFPA 1031, 4.2.1 and 4.2.1(A)(B)
IFSTA, *Fire Inspection and Code Enforcement,* 6th Edition, 2nd Printing, pages 33–34.
Answer: D

7. Reference: NFPA 1031, 4.2.2 and 4.2.2(A)(B)
IFSTA, *Fire Inspection and Code Enforcement,* 6th Edition, 1st Printing, pages 19–20.
Answer: B

8. Reference: NFPA 1031, 4.2.2 and 4.2.2(A)(B)
IFSTA, *Fire Inspection and Code Enforcement,* 6th Edition, 2nd Printing, page 20.
Answer: B

9. Reference: NFPA 1031, 4.2.3 and 4.2.3(A)(B)
IFSTA, *Fire Inspection and Code Enforcement,* 6th Edition, 2nd Printing, page 290.
Answer: C

10. Reference: NFPA 1031, 4.2.3 and 4.2.3(A)
IFSTA, *Fire Inspection and Code Enforcement,* 6th Edition, 2nd Printing, page 297.
Answer: D

11. Reference: NFPA 1031, 4.2.3, 4.2.3(A), 4.3.9, and 4.3.9(A)(B)
IFSTA, *Fire Inspection and Code Enforcement,* 6th Edition, 2nd Printing, pages 290–292.
Answer: D

12. Reference: NFPA 1031, 4.2.3 and 4.2.3(A)
IFSTA, *Fire Inspection and Code Enforcement,* 6th Edition, 2nd Printing, page 289.
Answer: C

13. Reference: NFPA 1031, 4.2.4 and 4.2.4(A)(B)
IFSTA, *Fire Inspection and Code Enforcement,* 6th Edition, 1st Printing, pages 14–15.
Answer: D

14. Reference: NFPA 1031, 4.2.5 and 4.2.5(A)(B)
IFSTA, *Fire Inspection and Code Enforcement,* 6th Edition, 1st Printing, page 379.
Answer: B

15. Reference: NFPA 1031, 4.2.5 and 4.2.5(A)(B)
IFSTA, *Fire Inspection and Code Enforcement,* 6th Edition, 2nd Printing, pages 16–17.
Answer: B

16. Reference: NFPA 1031, 4.2.6 and 4.2.6(A)(B)
IFSTA, *Fire Inspection and Code Enforcement,* 6th Edition, 2nd Printing, pages 13–14.
Answer: D

17. Reference: NFPA 1031, 4.2.6 and 4.2.6(A)(B)
IFSTA, *Fire Inspection and Code Enforcement,* 6th Edition, 2nd Printing, pages 13–14.
Answer: C

18. Reference: NFPA 1031, 4.3.10 and 4.3.10(A)(B)
IFSTA, *Fire Inspection and Code Enforcement,* 6th Edition, 1st Printing, page 153.
Answer: D

19. Reference: NFPA 1031, 4.3.10 and 4.3.10(A)(B)
IFSTA, *Fire Inspection and Code Enforcement,* 6th Edition, 1st Printing, page 152.
Answer: B

20. Reference: NFPA 1031, 4.3.10 and 4.3.10(A)(B)
IFSTA, *Fire Inspection and Code Enforcement,* 6th Edition, 2nd Printing, page 153.
Answer: D

21. Reference 1031, NFPA 4.3.11 and 4.3.11(A)(B)
IFSTA, *Fire Inspection and Code Enforcement,* 6th Edition, 1st Printing, page 136.
Answer: B

22. Reference 1031, NFPA 4.3.11 and 4.3.11(A)(B)
IFSTA, *Fire Inspection and Code Enforcement,* 6th Edition, 1st Printing, page 137.
Answer: D

23. Reference: NFPA 1031, 4.3.12 and 4.3.12(A)(B)
IFSTA, *Fire Inspection and Code Enforcement,* 6th Edition, 1st Printing, page 322.
Answer: A

24. Reference: NFPA 1031, 4.3.12 and 4.3.12(A)(B)
IFSTA, *Fire Inspection and Code Enforcement,* 6th Edition, 1st Printing, page 322.
Answer: D

25. Reference: NFPA 1031, 4.3.12 and 4.3.12(A)(B)
IFSTA, *Fire Inspection and Code Enforcement,* 6th Edition, 1st Printing, page 230.
Answer: A

26. Reference: NFPA 1031, 4.3.12 and 4.3.12(A)(B)
IFSTA, *Fire Inspection and Code Enforcement,* 6th Edition, 1st Printing, page 323.
Answer: C

27. Reference: NFPA 1031, 4.3.12 and 4.3.12(A)(B)
IFSTA, *Fire Inspection and Code Enforcement,* 6th Edition, 1st Printing, page 349.
Answer: B

28. Reference: NFPA 1031, 4.3.12 and 4.3.12(A)(B)
IFSTA, *Fire Inspection and Code Enforcement,* 6th Edition, 1st Printing, page 337.
Answer: C

29. Reference: NFPA 1031, 4.3.12 and 4.3.12(A)(B)
IFSTA, *Fire Inspection and Code Enforcement,* 6th Edition, 1st Printing, pages 332 and 335.
Answer: B

30. Reference: NFPA 1031, 4.3.12 and 4.3.12(A)(B)
IFSTA, *Fire Inspection and Code Enforcement,* 6th Edition, 1st Printing, pages 336–337.
Answer: C

31. Reference: NFPA 1031, 4.3.12 and 4.3.12(A)(B)
IFSTA, *Fire Inspection and Code Enforcement,* 6th Edition, 1st Printing, page 337.
Answer: C

32. Reference: NFPA 1031, 4.3.12 and 4.3.12(A)(B)
IFSTA, *Fire Inspection and Code Enforcement,* 6th Edition, 1st Printing, pages 325–326.
Answer: D

33. Reference: NFPA 1031, 4.3.1.2 and 4.3.12(A)(B)
IFSTA, *Fire Inspection and Code Enforcement,* 6th Edition, 1st Printing, page 332.
Answer: D

34. Reference: NFPA 1031, 4.3.12 and 4.3.12(A)(B)
IFSTA, *Fire Inspection and Code Enforcement,* 6th Edition, 1st Printing, page 325.
Answer: C

35. Reference: NFPA 1031, 4.3.12 and 4.3.12(A)(B)
IFSTA, *Fire Inspection and Code Enforcement,* 6th Edition, 1st Printing, page 327.
Answer: C

36. Reference: NFPA 1031, 4.3.12 and 4.3.12(A)(B)
IFSTA, *Fire Inspection and Code Enforcement,* 6th Edition, 1st Printing, page 350.
Answer: A

37. Reference: NFPA 1031, 4.3.12 and 4.3.12(A)(B)
IFSTA, *Fire Inspection and Code Enforcement,* 6th Edition, 1st Printing, page 351.
Answer: B

38. Reference: NFPA 1031, 4.3.13 and 4.3.13(A)(B)
IFSTA, *Fire Inspection and Code Enforcement,* 6th Edition, 1st Printing, page 99.
Answer: D

39. Reference: NFPA 1031, 4.3.13 and 4.3.13(A)(B)
IFSTA, *Fire Inspection and Code Enforcement,* 6th Edition, 1st Printing, page 357.
Answer: B

40. Reference: NFPA 1031, 4.3.13 and 4.3.13(A)(B)
IFSTA, *Fire Inspection and Code Enforcement,* 6th Edition, 1st Printing, page 356.
Answer: C

41. Reference: NFPA 1031, 4.3.14 and 4.3.14(A)(B)
IFSTA, *Fire Inspection and Code Enforcement,* 6th Edition, 1st Printing, page 234.
Answer: B

42. Reference: NFPA 1031, 4.3.14 and 4.3.14(A)(B)
IFSTA, *Essentials of Fire Fighting,* 4th Edition, 1st Printing, page 42.
Answer: B

43. Reference: NFPA 1031, 4.3.14 and 4.3.14(A)(B)
IFSTA, *Fire Inspection and Code Enforcement,* 6th Edition, 1st Printing, page 51.
Answer: B

44. Reference: NFPA 1031, 4.3.14 and 4.3.14(A)(B)
IFSTA, *Fire Inspection and Code Enforcement,* 6th Edition, 1st Printing, page 49.
Answer: C

45. Reference: NFPA 1031, 4.3.14 and 4.3.14(A)(B)
IFSTA, *Fire Inspection and Code Enforcement,* 6th Edition, 1st Printing, page 49.
Answer: A

46. Reference: NFPA 1031, 4.3.14 and 4.3.14(A)(B)
IFSTA, *Fire Inspection and Code Enforcement,* 6th Edition, 1st Printing, page 49.
Answer: B

47. Reference: NFPA 1031, 4.3.14 and 4.3.14(A)(B)
IFSTA, *Fire Inspection and Code Enforcement,* 6th Edition, 1st Printing, page 49.
Answer: B

48. Reference: NFPA 1031, 4.3.14 and 4.3.14(A)(B)
IFSTA, *Fire Inspection and Code Enforcement,* 6th Edition, 1st Printing, pages 53 and 56.
Answer: B

49. Reference: NFPA 1031, 4.3.14 and 4.3.14(A)(B)
IFSTA, *Fire Inspection and Code Enforcement,* 6th Edition, 1st Printing, page 56.
Answer: A

50. Reference: NFPA 1031, 4.3.14 and 4.3.14(A)(B)
IFSTA, *Fire Inspection and Code Enforcement,* 6th Edition, 1st Printing, page 50.
Answer: B

51. Reference: NFPA 1031, 4.3.14 and 4.3.14(A)(B)
IFSTA, *Fire Inspection and Code Enforcement,* 6th Edition, 1st Printing, page 54.
Answer: A

52. Reference: NFPA 1031, 4.3.14 and 4.3.14(A)(B)
IFSTA, *Fire Inspection and Code Enforcement,* 6th Edition, 1st Printing, page 48.
Answer: A

53. Reference: NFPA 1031, 4.3.14 and 4.3.14(A)(B)
IFSTA, *Fire Inspection and Code Enforcement,* 6th Edition, 1st Printing, page 48.
Answer: C

54. Reference: NFPA 1031, 4.3.14 and 4.3.14(A)(B)
IFSTA, *Fire Inspection and Code Enforcement,* 6th Edition, 1st Printing, page 48.
Answer: B

55. Reference: NFPA 1031, 4.3.14 and 4.3.14(A)(B)
IFSTA, *Fire Inspection and Code Enforcement,* 6th Edition, 1st Printing, page 56.
Answer: B

56. Reference: NFPA 1031, 4.3.14 and 4.3.14(A)(B)
IFSTA, *Fire Inspection and Code Enforcement,* 6th Edition, 1st Printing, page 100.
Answer: A

57. Reference: NFPA 1031, 4.3.14 and 4.3.14(A)(B)
IFSTA, *Fire Inspection and Code Enforcement,* 6th Edition, 1st Printing, page 51.
Answer: C

58. Reference: NFPA 1031, 4.3.14 and 4.3.14(A)(B)
IFSTA, *Fire Inspection and Code Enforcement,* 6th Edition, 1st Printing, page 54.
Answer: A

59. Reference: NFPA 1031, 4.3.14 and 4.3.14(A)(B)
IFSTA, *Fire Inspection and Code Enforcement,* 6th Edition, 1st Printing, page 51.
Answer: B

60. Reference: NFPA 1031, 4.3.14 and 4.3.14(A)
IFSTA, *Fire Inspection and Code Enforcement,* 6th Edition, 1st Printing, page 52.
Answer: D

61. Reference: NFPA 1031, 4.3.14 and 4.3.14(A)(B)
IFSTA, *Fire Inspection and Code Enforcement,* 6th Edition, 1st Printing, page 449.
Answer: A

62. Reference: NFPA 1031, 4.3.14 and 4.3.14(A)(B)
IFSTA, *Fire Inspection and Code Enforcement,* 6th Edition, 1st Printing, page 448.
Answer: D

63. Reference: NFPA 1031, 4.3.14 and 4.3.14(A)(B)
IFSTA, *Fire Inspection and Code Enforcement,* 6th Edition, 1st Printing, pages 48–49.
Answer: C

64. Reference: NFPA 1031, 4.3.14, 4.3.14(A)(B), 4.3.12, and 4.3.12(A)(B)
IFSTA, *Fire Inspection and Code Enforcement,* 6th Edition, 1st Printing, pages 322–323.
Answer: B

65. Reference: NFPA 1031, 4.3.14, 4.3.14(A)(B), 4.2.5, 4.2.5(A)(B), 4.3.1, and 4.3.1(A)(B)
IFSTA, *Fire Inspection and Code Enforcement,* 6th Edition, 1st Printing, pages 70–71.
Answer: B

66. Reference: NFPA 1031, 4.3.16 and 4.3.16(A)(B)
IFSTA, *Fire Inspection and Code Enforcement,* 6th Edition, 2nd Printing, page 209.
Answer: D

67. Reference: NFPA 1031, 4.3.16 and 4.3.16(A)(B)
IFSTA, *Fire Inspection and Code Enforcement,* 6th Edition, 2nd Printing, page 204.
Answer: D

68. Reference: NFPA 1031, 4.3.16 and 4.3.16(A)(B)
IFSTA, *Fire Inspection and Code Enforcement,* 6th Edition, 2nd Printing, page 208.
Answer: C

69. Reference: NFPA 1031, 4.3.16 and 4.3.16(A)
IFSTA, *Fire Inspection and Code Enforcement,* 6th Edition, 2nd Printing, page 208.
Answer: B

70. Reference: NFPA 1031, 4.3.2 and 4.3.2(A)
IFSTA, *Fire Inspection and Code Enforcement,* 6th Edition, 2nd Printing, page 145.
Answer: B

71. Reference: NFPA 1031, 4.3.3, 4.3.3(A)(B), 4.2.1, and 4.2.1(A)(B)
IFSTA, *Fire Inspection and Code Enforcement,* 6th Edition, 1st Printing, page 31.
Answer: D

72. Reference: NFPA 1031, 4.3.3, 4.3.3(A)(B), 4.2.5, and 4.2.5(A)(B)
IFSTA, *Fire Inspection and Code Enforcement,* 6th Edition, 1st Printing, page 66.
Answer: C

73. Reference: NFPA 1031, 4.3.3, 4.3.3(A)(B), 4.2.5, and 4.2.5(A)(B)
IFSTA, *Fire Inspection and Code Enforcement,* 6th Edition, 1st Printing, page 140.
Answer: B

74. Reference: NFPA 1031, 4.3.3, 4.3.3(A)(B), 4.2.5, and 4.2.5(A)(B)
IFSTA, *Fire Inspection and Code Enforcement,* 6th Edition, 1st Printing, page 138.
Answer: B

75. Reference: NFPA 1031, 4.3.3, 4.3.3(A)(B), 4.2.5, and 4.2.5(A)(B)
IFSTA, *Fire Inspection and Code Enforcement,* 6th Edition, 2nd Printing, page 151.
Answer: B

76. Reference: NFPA 1031, 4.3.3, 4.3.3(A)(B), 4.2.5, and 4.2.5(A)(B)
IFSTA, *Fire Inspection and Code Enforcement,* 6th Edition, 2nd Printing, page 138.
Answer: B

77. Reference: NFPA 1031, 4.3.3 and 4.3.3(A)
IFSTA, *Fire Inspection and Code Enforcement,* 6th Edition, 2nd Printing, page 143.
Answer: B

78. Reference: NFPA 1031, 4.3.4 and 4.3.4(A)(B)
IFSTA, *Fire Inspection and Code Enforcement,* 6th Edition, 1st Printing, pages 69–70.
Answer: A

79. Reference: NFPA 1031, 4.3.4 and 4.3.4(A)(B)
IFSTA, *Fire Inspection and Code Enforcement,* 6th Edition, 1st Printing, pages 69–70.
Answer: A

80. Reference: NFPA 1031, 4.3.5, 4.3.5(A)(B), 4.2.5, 4.2.5(A)(B), 4.3.4, and 4.3.4(A)(B)
IFSTA, *Fire Inspection and Code Enforcement,* 6th Edition, 1st Printing, page 180.
Answer: E

81. Reference: NFPA 1031, 4.3.5, 4.3.5(A)(B), 4.3.16, and 4.3.16(A)(B)
IFSTA, *Fire Inspection and Code Enforcement,* 6th Edition, 1st Printing, page 208.
Answer: B

82. Reference: NFPA 1031, 4.3.5, 4.3.5(A), 4.3.16, and 4.3.16(A)
IFSTA, *Essentials of Fire Fighting,* 4th Edition, 1st Printing, page 490.
Answer: C

83. Reference: NFPA 1031, 4.3.5, 4.3.5(A)(B), 4.3.16, and 4.3.16(A)(B)
IFSTA, *Fire Inspection and Code Enforcement,* 6th Edition, 1st Printing, page 160.
Answer: A

84. Reference: NFPA 1031, 4.3.5, 4.3.5(A)(B), 4.3.16, and 4.3.16(A)(B)
IFSTA, *Fire Inspection and Code Enforcement,* 6th Edition, 1st Printing, pages 160–161.
Answer: A

85. Reference: NFPA 1031, 4.3.5, 4.3.5(A), 4.3.16, and 4.3.16(A)(B)
IFSTA, *Fire Inspection and Code Enforcement,* 6th Edition, 1st Printing, page 159.
Answer: D

86. Reference: NFPA 1031, 4.3.5, 4.3.5(A)(B), 4.3.16, and 4.3.16(A)(B)
IFSTA, *Fire Inspection and Code Enforcement,* 6th Edition, 1st Printing, page 173.
Answer: A

87. Reference: NFPA 1031, 4.3.5, 4.3.5(A)(B), 4.3.16, and 4.3.16(A)(B)
IFSTA, *Fire Inspection and Code Enforcement,* 6th Edition, 1st Printing, pages 162–163.
Answer: C

88. Reference: NFPA 1031, 4.3.5, 4.3.5(A)(B), 4.3.16, and 4.3.16(A)(B)
IFSTA, *Fire Inspection and Code Enforcement,* 6th Edition, 1st Printing, page 163.
Answer: B

89. Reference: NFPA 1031, 4.3.5, 4.3.5(A)(B), 4.3.16, and 4.3.16(A)(B)
IFSTA, *Fire Inspection and Code Enforcement,* 6th Edition, 1st Printing, pages 160–161, p. 160 Figure 7.3.
Answer: D

90. Reference: NFPA 1031, 4.3.5, 4.3.5(A)(B), 4.3.16, and 4.3.16(A)(B)
IFSTA, *Fire Inspection and Code Enforcement,* 6th Edition, 1st Printing, page 206.
Answer: B

91. Reference: NFPA 1031, 4.3.5, 4.3.5(A)(B), 4.3.16, and 4.3.16(A)(B)
IFSTA, *Fire Inspection and Code Enforcement,* 6th Edition, 1st Printing, page 205.
Answer: C

92. Reference: NFPA 1031, 4.3.5, 4.3.5(A)(B), 4.3.16, and 4.3.16(A)(B)
IFSTA, *Fire Inspection and Code Enforcement,* 6th Edition, 1st Printing, pages 206–207.
Answer: D

93. Reference: NFPA 1031, 4.3.5, 4.3.5(A), 4.3.16, and 4.3.16(A)(B)
IFSTA, *Fire Inspection and Code Enforcement,* 6th Edition, 1st Printing, page 206.
Answer: D

94. Reference: NFPA 1031, 4.3.5, 4.3.5(A)(B), 4.3.16, and 4.3.16(A)(B)
IFSTA, *Fire Inspection and Code Enforcement,* 6th Edition, 1st Printing, page 217.
Answer: D

95. Reference: NFPA 1031, 4.3.5, 4.3.5(A)(B), 4.3.16, and 4.3.16(A)(B)
IFSTA, *Fire Inspection and Code Enforcement,* 6th Edition, 1st Printing, page 453.
Answer: B

96. Reference: NFPA 1031, 4.3.5, 4.3.5(A)(B), 4.3.16, and 4.3.16(A)(B)
IFSTA, *Fire Inspection and Code Enforcement,* 6th Edition, 1st Printing, page 208.
Answer: C

97. Reference: NFPA 1031, 4.3.5, 4.3.5(A)(B), 4.3.16, 4.3.16(A)(B), 4.2.1, and 4.2.1(A)(B)
IFSTA, *Fire Inspection and Code Enforcement,* 6th Edition, 1st Printing, page 167.
Answer: B

98. Reference: NFPA 1031, 4.3.5, 4.3.5(A)(B), 4.3.16, and 4.3.16(A)(B)
IFSTA, *Fire Inspection and Code Enforcement,* 6th Edition, 1st Printing, page 192.
Answer: D

99. Reference: NFPA 1031, 4.3.5, 4.3.5(A)(B), 4.3.16, and 4.3.16(A)(B)
IFSTA, *Fire Inspection and Code Enforcement,* 6th Edition, 1st Printing, pages 162–163.
Answer: C

100. Reference: NFPA 1031, 4.3.5, 4.3.5(A)(B), 4.3.16, and 4.3.16(A)(B)
IFSTA, *Fire Inspection and Code Enforcement,* 6th Edition, 1st Printing, page 162.
Answer: C

101. Reference: NFPA 1031, 4.3.5 and 4.3.5(A)(B)
IFSTA, *Fire Inspection and Code Enforcement,* 6th Edition, 1st Printing, page 249.
Answer: C

102. Reference: NFPA 1031, 4.3.5, and 4.3.5(A)(B)
IFSTA, *Fire Inspection and Code Enforcement,* 6th Edition, 1st Printing, pages 187–188.
Answer: C

103. Reference: NFPA 1031, 4.3.5 and 4.3.5(A)(B)
IFSTA, *Fire Inspection and Code Enforcement,* 6th Edition, 1st Printing, pages 186–187
Answer: A

104. Reference: NFPA 1031, 4.3.5, 4.3.5(A)(B), 4.3.16, and 4.3.16(A)(B)
IFSTA, *Fire Inspection and Code Enforcement,* 6th Edition, 1st Printing, page 169.
Answer: C

105. Reference: NFPA 1031, 4.3.5, 4.3.5(A)(B), 4.3.16, and 4.3.16(A)(B)
IFSTA, *Fire Inspection and Code Enforcement,* 6th Edition, 1st Printing, page 167, Figure 7.20.
Answer: B

106. Reference: NFPA 1031, 4.3.5, 4.3.5(A)(B), 4.3.16, and 4.3.16(A)(B)
IFSTA, *Fire Inspection and Code Enforcement,* 6th Edition, 1st Printing, pages 176–178.
Answer: B

107. Reference: NFPA 1031, 4.3.5, 4.3.5(A)(B), 4.3.16, and 4.3.16(A)(B)
IFSTA, *Fire Inspection and Code Enforcement,* 6th Edition, 1st Printing, page 174.
Answer: B

108. Reference: NFPA 1031, 4.3.5, 4.3.5(A)(B), 4.3.16, and 4.3.16(A)(B)
IFSTA, *Fire Inspection and Code Enforcement,* 6th Edition, 1st Printing, pages 206–207, Figure 7.103.
Answer: A

109. Reference: NFPA 1031, 4.3.5, 4.3.5(A)(B), 4.3.16, and 4.3.16(A)(B)
IFSTA, *Fire Inspection and Code Enforcement,* 6th Edition, 1st Printing, page 207, Figure 7.103.
Answer: C

110. Reference: NFPA 1031, 4.3.5, 4.3.5(A)(B), 4.3.16, and 4.3.16(A)(B)
IFSTA, *Fire Inspection and Code Enforcement,* 6th Edition, 1st Printing, page 210.
Answer: B

111. Reference: NFPA 1031, 4.3.5, 4.3.5(A)(B), 4.3.16, and 4.3.16(A)(B)
IFSTA, *Fire Inspection and Code Enforcement,* 6th Edition, 1st Printing, page 205.
Answer: B

112. Reference: NFPA 1031, 4.3.5, 4.3.5(A)(B), 4.3.16, and 4.3.16(A)(B)
IFSTA, *Fire Inspection and Code Enforcement,* 6th Edition, 1st Printing, page 163.
Answer: B

113. Reference: NFPA 1031, 4.3.5, 4.3.5(A)(B), 4.3.16, and 4.3.16(A)(B)
IFSTA, *Fire Inspection and Code Enforcement,* 6th Edition, 1st Printing, page 161, Figure 7.4.
Answer: C

114. Reference: NFPA 1031, 4.3.5, 4.3.5(A), 4.2.5, and 4.2.5(A)(B)
IFSTA, *Fire Inspection and Code Enforcement,* 6th Edition, 2nd Printing, page 254.
Answer: B

115. Reference: NFPA 1031, 4.3.5, 4.3.5(A), 4.3.16, and 4.3.16(A)(B)
IFSTA, *Fire Inspection and Code Enforcement,* 6th Edition, 2nd Printing, page 163.
Answer: A

116. Reference: NFPA 1031, 4.3.6 and 4.3.6(A)(B)
IFSTA, *Fire Inspection and Code Enforcement,* 6th Edition, 1st Printing, pages 306–307.
Answer: D

117. Reference: NFPA 1031, 4.3.6 and 4.3.6(A)(B)
IFSTA, *Fire Inspection and Code Enforcement,* 6th Edition, 1st Printing, page 270.
Answer: D

118. Reference: NFPA 1031, 4.3.6 and 4.3.6(A)(B)
IFSTA, *Fire Inspection and Code Enforcement,* 6th Edition, 1st Printing, page 270.
Answer: C

119. Reference: NFPA 1031, 4.3.6 and 4.3.6(A)(B)
IFSTA, *Fire Inspection and Code Enforcement,* 6th Edition, 2nd Printing, page 90.
Answer: D

120. Reference: NFPA 1031, 4.3.7 and 4.3.7(A)
IFSTA, *Fire Inspection and Code Enforcement,* 6th Edition, 1st Printing, page 230.
Answer: C

121. Reference: NFPA 1031, 4.3.7 and 4.3.7(A)(B)
IFSTA, *Fire Inspection and Code Enforcement,* 6th Edition, 1st Printing, page 232.
Answer: B

122. Reference: NFPA 1031, 4.3.7 and 4.3.7(A)(B)
IFSTA, *Fire Inspection and Code Enforcement,* 6th Edition, 1st Printing, page 244.
Answer: B

123. Reference: NFPA 1031, 4.3.7 and 4.3.7(A)(B)
IFSTA, *Fire Inspection and Code Enforcement,* 6th Edition, 1st Printing, pages 231–233.
Answer: A

124. Reference: NFPA 1031, 4.3.7 and 4.3.7(A)(B)
IFSTA, *Fire Inspection and Code Enforcement,* 6th Edition, 1st Printing, page 234, Figure 8.11.
Answer: D

125. Reference: NFPA 1031, 4.3.7 and 4.3.7(A)(B)
IFSTA, *Fire Inspection and Code Enforcement,* 6th Edition, 1st Printing, page 237.
Answer: D

126. Reference: NFPA 1031, 4.3.7 and 4.3.7(A)(B)
IFSTA, *Fire Inspection and Code Enforcement,* 6th Edition, 1st Printing, pages 230, 232, and 234.
Answer: A

127. Reference: NFPA 1031, 4.3.7 and 4.3.7(A)(B)
IFSTA, *Fire Inspection and Code Enforcement,* 6th Edition, 1st Printing, page 234, Figure 8.11.
Answer: B

128. Reference: NFPA 1031, 4.3.7 and 4.3.7(A)(B)
IFSTA, *Fire Inspection and Code Enforcement,* 6th Edition, 1st Printing, pages 231 and 234.
Answer: D

129. Reference: NFPA 1031, 4.3.7 and 4.3.7(A)(B)
IFSTA, *Fire Inspection and Code Enforcement,* 6th Edition, 1st Printing, page 238.
Answer: B

130. Reference: NFPA 1031, 4.3.7 and 4.3.7(A)(B)
IFSTA, *Fire Inspection and Code Enforcement,* 6th Edition, 1st Printing, page 233, Figure 8.10.
Answer: B

131. Reference: NFPA 1031, 4.3.7 and 4.3.7(A)(B)
Delmar, *Firefighter's Handbook,* 1st Edition, 1st Printing, page 174.
Answer: B

132. Reference: NFPA 1031, 4.3.7 and 4.3.7(A)(B)
IFSTA, *Fire Inspection and Code Enforcement,* 6th Edition, 1st Printing, page 246.
Answer: C

133. Reference: NFPA 1031, 4.3.7 and 4.3.7(A)(B)
IFSTA, *Fire Inspection and Code Enforcement,* 6th Edition, 1st Printing, page 248.
Answer: A

134. Reference: NFPA 1031, 4.3.7 and 4.3.7(A)(B)
IFSTA, *Fire Inspection and Code Enforcement,* 6th Edition, 1st Printing, page 248.
Answer: A

135. Reference: NFPA 1031, 4.3.7 and 4.3.7(A)(B)
IFSTA, *Fire Inspection and Code Enforcement,* 6th Edition, 1st Printing, pages 231–232.
Answer: C

136. Reference: NFPA 1031, 4.3.7 and 4.3.7(A)(B)
IFSTA, *Fire Inspection and Code Enforcement,* 6th Edition, 1st Printing, page 231, Figure 8.5.
Answer: D

137. Reference: NFPA 1031, 4.3.7 and 4.3.7(A)(B)
IFSTA, *Fire Inspection and Code Enforcement,* 6th Edition, 1st Printing, pages 230–231.
Answer: A

138. Reference: NFPA 1031, 4.3.7 and 4.3.7(A)(B)
IFSTA, *Fire Inspection and Code Enforcement,* 6th Edition, 1st Printing, page 231, Figure 8.5.
Answer: A

139. Reference: NFPA 1031, 4.3.7 and 4.3.7(A)(B)
IFSTA, *Fire Inspection and Code Enforcement,* 6th Edition, 1st Printing, page 230, Figure 8.2.
Answer: A

140. Reference: NFPA 1031, 4.3.7 and 4.3.7(A)(B)
IFSTA, *Fire Inspection and Code Enforcement,* 6th Edition, 1st Printing, page 230, Figure 8.3.
Answer: B

141. Reference: NFPA 1031, 4.3.7 and 4.3.7(A)(B)
IFSTA, *Fire Inspection and Code Enforcement,* 6th Edition, 2nd Printing, page 248.
Answer: C

142. Reference: NFPA 1031, 4.3.8 and 4.3.8(A)(B)
IFSTA, *Fire Inspection and Code Enforcement,* 6th Edition, 1st Printing, page 453.
Answer: A

143. Reference: NFPA 1031, 4.3.8 and 4.3.8(A)(B)
IFSTA, *Fire Inspection and Code Enforcement,* 6th Edition, 1st Printing, pages 77–78.
Answer: B

144. Reference: NFPA 1031, 4.3.8 and 4.3.8(A)(B)
IFSTA, *Fire Inspection and Code Enforcement,* 6th Edition, 1st Printing, pages 50–51.
Answer: D

145. Reference: NFPA 1031, 4.3.8 and 4.3.8(A)
IFSTA, *Fire Inspection and Code Enforcement,* 6th Edition, 1st Printing, pages 87–90.
Answer: B

146. Reference: NFPA 1031, 4.3.8 and 4.3.8(A)
IFSTA, *Fire Inspection and Code Enforcement,* 6th Edition, 1st Printing, pages 93–94.
Answer: C

147. Reference: NFPA 1031, 4.3.8 and 4.3.8(A)(B)
IFSTA, *Fire Inspection and Code Enforcement,* 6th Edition, 1st Printing, page 90.
Answer: A

148. Reference: NFPA 1031, 4.3.8, 4.3.8(A)(B), 4.2.1, and 4.2.1(A)(B)
IFSTA, *Fire Inspection and Code Enforcement,* 6th Edition, 1st Printing, page 40.
Answer: A

149. Reference: NFPA 1031, 4.3.9, 4.3.9(A)(B), 4.3.8, and 4.3.8(A)(B)
IFSTA, *Fire Inspection and Code Enforcement,* 6th Edition, 1st Printing, pages 69 and 90.
Answer: D

150. Reference: NFPA 1031, 4.3.9, 4.3.9(A)(B), 4.3.10, 4.3.10(A)(B), 4.3.14, and 4.3.14(A)(B)
IFSTA, *Fire Inspection and Code Enforcement,* 6th Edition, 1st Printing, page 289.
Answer: D

APPENDIX B

Examination II-1 Answer Key

Directions

Follow these steps carefully for completing the feedback part of the Systematic Approach to Examination Preparation (SAEP):

1. After entering your scores, look up the answers for the examination items you missed as well as those you guessed, even if you guessed correctly. If you are guessing, it means the answer isn't perfectly clear. This process will make you as knowledgeable as possible.
2. Enter the number of missed and guessed examination items in the blank on your Personal Progress Plotter.
3. Highlight the answer in the reference materials, and read the paragraph preceding and the paragraph following the one in which the correct answer is located. Enter the paragraph number and page number next to the guessed or missed examination item on your examination. Count any part of a paragraph at the beginning of the page as one paragraph until you reach the paragraph containing your highlighted answer. This step will help you locate and review your missed and guessed examination items later in the process. This step is essential to learning the material in context and by association. These learning techniques (context/association) are the very backbone of the SAEP approach.
4. Once you have completed the feedback part, you may proceed to the next examination.

1. Reference: NFPA 1031, 5.2
IFSTA, *Fire Inspection and Code Enforcement,* 6th Edition, pages 13–14.
Answer: B

2. Reference: NFPA 1031, 5.2
IFSTA, *Fire Inspection and Code Enforcement,* 6th Edition, page 13.
Answer: D

3. Reference: NFPA 1031, 5.2.1, 5.2.1(A)(B), 5.2.4, and 5.2.4(A)(B)
IFSTA, *Fire Inspection and Code Enforcement,* 6th Edition, pages 19–20.
Answer: B

4. Reference: NFPA 1031, 5.2.2 and 5.2.2(A)(B)
IFSTA, *Fire Inspection and Code Enforcement,* 6th Edition, page 289.
Answer: A

5. Reference: NFPA 1031, 5.2.2 and 5.2.2(A)(B)
IFSTA, *Fire Inspection and Code Enforcement,* 6th Edition, page 289.
Answer: B

6. Reference: NFPA 1031, 5.2.3 and 5.2.3(A)(B)
IFSTA, *Fire Inspection and Code Enforcement,* 6th Edition, pages 14–15.
Answer: C

7. Reference: NFPA 1031, 5.2.3 and 5.2.3(A)(B)
IFSTA, *Fire Inspection and Code Enforcement,* 6th Edition, pages 14–15.
Answer: B

8. Reference: NFPA 1031, 5.2.4 and 5.2.4(A)(B)
IFSTA, *Fire Inspection and Code Enforcement,* 6th Edition, pages 11–12.
Answer: C

9. Reference: NFPA 1031, 5.2.4 and 5.2.4(A)(B)
IFSTA, *Fire Inspection and Code Enforcement,* 6th Edition, pages 11–12.
Answer: A

10. Reference: NFPA 1031, 5.2.5, 5.2.5(A)(B), 5.2.3, and 5.2.3(A)(B)
IFSTA, *Fire Inspection and Code Enforcement,* 6th Edition, page 36.
Answer: D

11. Reference: NFPA 1031, 5.2.5 and 5.2.5(A)(B)
IFSTA, *Fire Inspection and Code Enforcement,* 6th Edition, page 36.
Answer: A

12. Reference: NFPA 1031, 5.2.5 and 5.2.5(A)(B)
IFSTA, *Fire Inspection and Code Enforcement,* 6th Edition, page 36.
Answer: B

13. Reference: NFPA 1031, 5.2.5, 5.2.5(A)(B), 5.2.3, and 5.2.3(A)(B)
IFSTA, *Fire Inspection and Code Enforcement,* 6th Edition, page 36.
Answer: B

14. Reference: NFPA 1031, 5.3.1 and 5.3.1(A)(B)
IFSTA, *Fire Inspection and Code Enforcement,* 6th Edition, page 145.
Delmar, *Fire Prevention, Inspection, and Code Enforcement,* 2nd Edition, page 103.
Answer: A

15. Reference: NFPA 1031, 5.3.1, 5.3.1(A)(B), 5.3.2, and 5.3.2(A)(B)
IFSTA, *Fire Inspection and Code Enforcement,* 6th Edition, page 147.
Answer: C

16. Reference: NFPA 1031, 5.3.1, 5.3.1(A)(B), 5.3.2, and 5.3.2(A)(B)
IFSTA, *Fire Inspection and Code Enforcement,* 6th Edition, page 146.
Answer: A

17. Reference: NFPA 1031, 5.3.10 and 5.3.10(A)(B)
IFSTA, *Fire Inspection and Code Enforcement,* 6th Edition, page 71.
Answer: B

18. Reference: NFPA 1031, 5.3.10 and 5.3.10(A)(B)
IFSTA, *Fire Inspection and Code Enforcement,* 6th Edition, page 59.
Delmar, *Fire Prevention, Inspection, and Code Enforcement,* 2nd Edition, page 123.
Answer: E

19. Reference: NFPA 1031, 5.3.10 and 5.3.10(A)(B)
IFSTA, *Fire Inspection and Code Enforcement,* 6th Edition, pages 59-60.
Delmar, *Fire Prevention, Inspection, and Code Enforcement,* 2nd Edition, page 123.
Answer: D

20. Reference: NFPA 1031, 5.3.11 and 5.3.11(A)(B)
IFSTA, *Fire Inspection and Code Enforcement,* 6th Edition, page 136.
Delmar, *Fire Prevention, Inspection, and Code Enforcement,* 2nd Edition, page 147.
Answer: C

21. Reference: NFPA 1031, 5.3.11 and 5.3.11(A)(B)
IFSTA, *Fire Inspection and Code Enforcement,* 6th Edition, page 136.
Delmar, *Fire Prevention, Inspection, and Code Enforcement,* 2nd Edition, page 147.
Answer: A

22. Reference: NFPA 1031, 5.3.11 and 5.3.11(A)(B)
IFSTA, *Fire Inspection and Code Enforcement,* 6th Edition, page 136.
Delmar, *Fire Prevention, Inspection, and Code Enforcement,* 2nd Edition, page 123.
Answer: D

23. Reference: NFPA 1031, 5.3.12, 5.3.12(A)(B), 5.3.6, and 5.3.6(A)(B)
IFSTA, *Fire Inspection and Code Enforcement,* 6th Edition, page 75.
Answer: B

24. Reference: NFPA 1031, 5.3.12, 5.3.12(A)(B), 5.3.7, 5.3.7(A)(B), 5.3.5, and 5.3.5(A)(B)
IFSTA, *Fire Inspection and Code Enforcement,* 6th Edition, pages 141 and 301.
Answer: C

25. Reference: NFPA 1031, 5.3.2, 5.3.2(A)(B), 5.4.1, 5.4.1(A)(B), 5.4.2, 5.4.2(A)(B), 5.4.5, 5.4.5(A)(B), 5.3.12, and 5.3.12(A)(B)
IFSTA, *Fire Inspection and Code Enforcement,* 6th Edition, page 115.
Answer: D

26. Reference: NFPA 1031, 5.3.3, 5.3.3(A)(B), 5.3.5, 5.3.5(A)(B), 5.4.4, and 5.4.4(A)(B)
IFSTA, *Fire Inspection and Code Enforcement,* 6th Edition, page 141.
Delmar, *Fire Prevention, Inspection, and Code Enforcement,* 2nd Edition, page 111.
Answer: B

27. Reference: NFPA 1031, 5.3.3, 5.3.3(A)(B), 5.3.5, and 5.3.5(A)(B)
IFSTA, *Fire Inspection and Code Enforcement,* 6th Edition, page 140.
Answer: A

28. Reference: NFPA 1031, 5.3.4, 5.3.4(A)(B), 5.4.3, 5.4.3(A), 5.3.1, and 5.3.1(A)(B)
IFSTA, *Fire Inspection and Code Enforcement,* 6th Edition, page 60.
Answer: C

29. Reference: NFPA 1031, 5.3.4, 5.3.4(A)(B), 5.3.3, and 5.3.3(A)(B)
IFSTA, *Fire Inspection and Code Enforcement,* 6th Edition, page 67.
Answer: A

30. Reference: NFPA 1031, 5.3.4, 5.3.4(A)(B), 5.3.8, and 5.3.8(A)(B)
IFSTA, *Fire Inspection and Code Enforcement,* 6th Edition, page 234.
Answer: A

31. Reference: NFPA 1031, 5.3.4 and 5.3.4(A)(B)
IFSTA, *Fire Inspection and Code Enforcement,* 6th Edition, page 249.
Answer: C

32. Reference: NFPA 1031, 5.3.4 and 5.3.4(A)(B)
IFSTA, *Fire Inspection and Code Enforcement,* 6th Edition, pages 249–257.
Answer: A

33. Reference: NFPA 1031, 5.3.4 and 5.3.4(A)(B)
IFSTA, *Fire Inspection and Code Enforcement,* 6th Edition, page 164.
Delmar, *Fire Prevention, Inspection, and Code Enforcement,* 2nd Edition, page 152.
Answer: D

34. Reference: NFPA 1031, 5.3.4, 5.3.4(A)(B), 5.3.5, 5.3.5(A)(B), 5.3.2, and 5.3.2(A)(B)
IFSTA, *Fire Inspection and Code Enforcement,* 6th Edition, page 65.
Answer: E

35. Reference: NFPA 1031, 5.3.4, 5.3.4(A)(B), 5.3.5, 5.3.5(A)(B), 5.4.5, and 5.4.5(A)(B)
IFSTA, *Fire Inspection and Code Enforcement,* 6th Edition, page 65.
Answer: E

36. Reference: NFPA 1031, 5.3.4, 5.3.4(A)(B), 5.4.3, and 5.4.3(A)(B)
IFSTA, *Fire Inspection and Code Enforcement,* 6th Edition, page 452.
Answer: B

37. Reference: NFPA 1031, 5.3.4, 5.3.4(A)(B), 5.4.3, and 5.4.3(A)(B)
IFSTA, *Fire Inspection and Code Enforcement,* 6th Edition, page 164.
Answer: D

38. Reference: NFPA 1031, 5.3.4, 5.3.4(A)(B), 5.4.3, and 5.4.3(A)(B)
IFSTA, *Fire Inspection and Code Enforcement,* 6th Edition, page 210.
Delmar, *Fire Prevention, Inspection, and Code Enforcement,* 2nd Edition, page 164.
Answer: B

39. Reference: NFPA 1031, 5.3.4 and 5.3.4(A)(B)
IFSTA, *Fire Inspection and Code Enforcement,* 6th Edition, page 210.
Answer: C

40. Reference: NFPA 1031, 5.3.4 and 5.3.4(A)(B)
IFSTA, *Fire Inspection and Code Enforcement,* 6th Edition, page 210.
Answer: B

41. Reference: NFPA 1031, 5.3.4, 5.3.4(A)(B), 5.4.3, and 5.4.3(A)(B)
IFSTA, *Fire Inspection and Code Enforcement,* 6th Edition, page 206.
Answer: D

42. Reference: NFPA 1031, 5.3.4, 5.3.4(A)(B), 5.4.1, and 5.4.1(A)(B)
IFSTA, *Fire Inspection and Code Enforcement,* 6th Edition, page 302.
Answer: A

43. Reference: NFPA 1031, 5.3.5 and 5.3.5(A)(B)
IFSTA, *Fire Inspection and Code Enforcement,* 6th Edition, page 139.
Answer: A

44. Reference: NFPA 1031, 5.3.5, 5.3.5(A)(B), 5.4.4, and 5.4.4(A)(B)
IFSTA, *Fire Inspection and Code Enforcement,* 6th Edition, page 138.
Answer: A

45. Reference: NFPA 1031, 5.3.5, 5.3.5(A)(B), 5.3.4, and 5.3.4(A)(B)
IFSTA, *Fire Inspection and Code Enforcement,* 6th Edition, page 140.
Answer: C

46. Reference: NFPA 1031, 5.3.5, 5.3.5(A)(B), 5.3.4, and 5.3.4(A)(B)
IFSTA, *Fire Inspection and Code Enforcement,* 6th Edition, page 144.
Answer: A

47. Reference: NFPA 1031, 5.3.5, 5.3.5(A)(B), 5.4.4, and 5.4.4(A)(B)
IFSTA, *Fire Inspection and Code Enforcement,* 6th Edition, page 144.
Delmar, *Fire Prevention, Inspection, and Code Enforcement,* 2nd Edition, page 150.
Answer: A

48. Reference: NFPA 1031, 5.3.6 and 5.3.6(A)(B)
IFSTA, *Fire Inspection and Code Enforcement,* 6th Edition, page 93.
Answer: D

49. Reference: NFPA 1031, 5.3.6 and 5.3.6(A)(B)
IFSTA, *Fire Inspection and Code Enforcement,* 6th Edition, page 75.
Answer: A

50. Reference: NFPA 1031, 5.3.7 and 5.3.7(A)(B)
IFSTA, *Fire Inspection and Code Enforcement,* 6th Edition, page 152.
Delmar, *Fire Prevention, Inspection, and Code Enforcement,* 2nd Edition, pages 180–181.
Answer: B

51. Reference: NFPA 1031, 5.3.7 and 5.3.7(A)(B)
IFSTA, *Fire Inspection and Code Enforcement,* 6th Edition, page 153.
Delmar, *Fire Prevention, Inspection, and Code Enforcement,* 2nd Edition, page 181.
Answer: A

52. Reference: NFPA 1031, 5.3.8, 5.3.8(A)(B), 5.3.9, and 5.3.9(A)(B)
IFSTA, *Fire Inspection and Code Enforcement,* 6th Edition, page 350.
Answer: C

53. Reference: NFPA 1031, 5.3.8, 5.3.8(A)(B), 5.3.9, and 5.3.9(A)(B)
IFSTA, *Fire Inspection and Code Enforcement,* 6th Edition, page 354.
Answer: A

54. Reference: NFPA 1031, 5.3.8 and 5.3.8(A)(B)
IFSTA, *Fire Inspection and Code Enforcement,* 6th Edition, pages 323–324.
Answer: C

55. Reference: NFPA 1031, 5.3.8 and 5.3.8(A)(B)
IFSTA, *Fire Inspection and Code Enforcement,* 6th Edition, pages 325–326.
Delmar, *Fire Prevention, Inspection, and Code Enforcement,* 2nd Edition, pages 202–203.
Answer: D

56. Reference: NFPA 1031, 5.3.9 and 5.3.9(A)(B)
IFSTA, *Fire Inspection and Code Enforcement,* 6th Edition, page 347.
Answer: D

57. Reference: NFPA 1031, 5.3.9, 5.3.9(A)(B), 5.3.8, and 5.3.8(A)(B)
IFSTA, *Fire Inspection and Code Enforcement,* 6th Edition, page 317.
Delmar, *Fire Prevention, Inspection, and Code Enforcement,* 2nd Edition, page 194.
Answer: A

58. Reference: NFPA 1031, 5.4.1, 5.4.1(A)(B), 5.3.2, and 5.3.2(A)(B)
IFSTA, *Fire Inspection and Code Enforcement,* 6th Edition, page 131.
Answer: A

59. Reference: NFPA 1031, 5.4.1, and 5.4.1(A)(B)
IFSTA, *Fire Inspection and Code Enforcement,* 6th Edition, page 122.
Answer: B

60. Reference: NFPA 1031, 5.4.2, and 5.4.2(A)(B), 5.3.5, 5.3.5(A)(B), 5.3.1, and 5.3.1(A)(B)
IFSTA, *Fire Inspection and Code Enforcement,* 6th Edition, page 150.
Delmar, *Fire Prevention, Inspection, and Code Enforcement,* 2nd Edition, page 110.
Answer: C

61. Reference: NFPA 1031, 5.4.2, 5.4.2(A)(B) 5.3.1, and 5.3.1(A)(B)
IFSTA, *Fire Inspection and Code Enforcement,* 6th Edition, page 147.
Delmar, *Fire Prevention, Inspection, and Code Enforcement,* 2nd Edition, pages 103–104.
Answer: A

62. Reference: NFPA 1031, 5.4.2, 5.4.2(A)(B), 5.3.1, and 5.3.1(A)(B)
IFSTA, *Fire Inspection and Code Enforcement,* 6th Edition, page 145.
Delmar, *Fire Prevention, Inspection, and Code Enforcement,* 2nd Edition, pages 103–104.
Answer: C

63. Reference: NFPA 1031, 5.4.3, 5.4.3(A)(B), 5.3.4, and 5.3.4(A)(B)
IFSTA, *Fire Inspection and Code Enforcement,* 6th Edition, page 217.
Delmar, *Fire Prevention, Inspection, and Code Enforcement,* 2nd Edition, page 164.
Answer: C

64. Reference: NFPA 1031, 5.4.3, 5.4.3(A)(B), 5.3.4, and 5.3.4(A)(B)
IFSTA, *Fire Inspection and Code Enforcement,* 6th Edition, page 205.
Answer: B

65. Reference: NFPA 1031, 5.4.3, 5.4.3(A)(B), 5.3.4, and 5.3.4(A)(B)
IFSTA, *Fire Inspection and Code Enforcement,* 6th Edition, page 226.
Answer: B

66. Reference: NFPA 1031, 5.4.3 and 5.4.3(A)(B)
IFSTA, *Fire Inspection and Code Enforcement,* 6th Edition, page 207.
Answer: C

67. Reference: NFPA 1031, 5.4.4, 5.4.4(A)(B), 5.3.5, 5.3.5(A)(B)
IFSTA, *Fire Inspection and Code Enforcement,* 6th Edition, page 142.
Answer: D

68. Reference: NFPA 1031, 5.4.5 and 5.4.5(A)(B)
IFSTA, *Fire Inspection and Code Enforcement,* 6th Edition, page 67.
Delmar, *Fire Prevention, Inspection, and Code Enforcement,* 2nd Edition, page 69.
Answer: D

69. Reference: NFPA 1031, 5.4.5 and 5.4.5(A)(B)
IFSTA, *Fire Inspection and Code Enforcement,* 6th Edition, page 58.
Answer: B

70. Reference: NFPA 1031, 5.4.5, 5.4.5(A)(B), 5.3.5, and 5.3.5(A)(B)
IFSTA, *Fire Inspection and Code Enforcement,* 6th Edition, page 116.
Delmar, *Fire Prevention, Inspection, and Code Enforcement,* 2nd Edition, page 51.
Answer: C

71. Reference: NFPA 1031, 5.4.5, 5.4.5(A)(B), 5.3.3, and 5.3.3(A)(B)
IFSTA, *Fire Inspection and Code Enforcement,* 6th Edition, pages 115–116.
Answer: A

72. Reference: NFPA 1031, 5.4.5, 5.4.5(A)(B), 5.3.3, and 5.3.3(A)(B)
IFSTA, *Fire Inspection and Code Enforcement,* 6th Edition, page 64.
Delmar, *Fire Prevention, Inspection, and Code Enforcement,* 2nd Edition, page 67.
Answer: C

73. Reference: NFPA 1031, 5.4.5, 5.4.5(A)(B), 5.3.3, and 5.3.3(A)(B)
IFSTA, *Fire Inspection and Code Enforcement,* 6th Edition, page 116.
Answer: B

74. Reference: NFPA 1031, 5.4.5, 5.4.5(A)(B), 5.3.3, and 5.3.3(A)(B)
IFSTA, *Fire Inspection and Code Enforcement,* 6th Edition, pages 63–64.
Answer: C

75. Reference: NFPA 1031, 5.4.5 and 5.4.5(A)(B)
IFSTA, *Fire Inspection and Code Enforcement,* 6th Edition, page 115.
Answer: C

Examination II-2 Answer Key

Directions

Follow these steps carefully for completing the feedback part of the Systematic Approach to Examination Preparation (SAEP):

1. After entering your scores, look up the answers for the examination items you missed as well as those you guessed, even if you guessed correctly. If you are guessing, it means the answer isn't perfectly clear. This process will make you as knowledgeable as possible.
2. Enter the number of missed and guessed examination items in the blank on your Personal Progress Plotter.
3. Highlight the answer in the reference materials, and then read the paragraph preceding and the paragraph following the one in which the correct answer is located. Enter the paragraph number and page number next to the guessed or missed examination item on your examination. Count any part of a paragraph at the beginning of the page as one paragraph until you reach the paragraph containing your highlighted answer. This step will help you locate and review your missed and guessed examination items later in the process. This step is essential to learning the material in context and by association. These learning techniques (context/association) are the very backbone of the SAEP approach.
4. Once you have completed the feedback part, you may proceed to the next examination.

1. Reference: NFPA 1031, 5.2, 5.2.3, and 5.2.3(A)(B)
   IFSTA, *Fire Inspection and Code Enforcement,* 6th Edition, pages 9–10.
   Answer: A

2. Reference: NFPA 1031, 5.2
   IFSTA, *Fire Inspection and Code Enforcement,* 6th Edition, page 14.
   Answer: C

3. Reference: NFPA 1031, 5.2.1 and 5.2.1(A)(B)
   IFSTA, *Fire Inspection and Code Enforcement,* 6th Edition, page 20.
   Answer: B

4. Reference: NFPA 1031, 5.2.1 and 5.2.1(A)(B)
   IFSTA, *Fire Inspection and Code Enforcement,* 6th Edition, page 20.
   Answer: D

5. Reference: NFPA 1031, 5.2.2 and 5.2.2(A)(B)
   IFSTA, *Fire Inspection and Code Enforcement,* 6th Edition, page 289.
   Answer: C

6. Reference: NFPA 1031, 5.2.2 and 5.2.2(A)(B)
   IFSTA, *Fire Inspection and Code Enforcement,* 6th Edition, page 289.
   Answer: D

7. Reference: NFPA 1031, 5.2.3 and 5.2.3(A)(B)
IFSTA, *Fire Inspection and Code Enforcement,* 6th Edition, pages 14–15.
Answer: D

8. Reference: NFPA 1031, 5.2.3 and 5.2.3(A)(B)
IFSTA, *Fire Inspection and Code Enforcement,* 6th Edition, pages 14–15.
Answer: A

9. Reference: NFPA 1031, 5.2.4 and 5.2.4(A)(B)
IFSTA, *Fire Inspection and Code Enforcement,* 6th Edition, pages 11–12.
Answer: D

10. Reference: NFPA 1031, 5.2.5 and 5.2.5(A)(B)
IFSTA, *Fire Inspection and Code Enforcement,* 6th Edition, page 36.
Answer: C

11. Reference: NFPA 1031, 5.2.5 and 5.2.5(A)(B)
IFSTA, *Fire Inspection and Code Enforcement,* 6th Edition, page 36.
Answer: C

12. Reference: NFPA 1031, 5.2.5, 5.2.5(A)(B), 5.2.4, and 5.2.4(A)(B)
IFSTA, *Fire Inspection and Code Enforcement,* 6th Edition, pages 379–380.
Answer: D

13. Reference: NFPA 1031, 5.3.1, 5.3.1(A)(B), 5.3.2, and 5.3.2(A)(B)
IFSTA, *Fire Inspection and Code Enforcement,* 6th Edition, page 147.
Answer: D

14. Reference: NFPA 1031, 5.3.1, 5.3.1(A)(B), 5.3.3, and 5.3.3(A)(B)
IFSTA, *Fire Inspection and Code Enforcement,* 6th Edition, pages 146–150.
Answer: A

15. Reference: NFPA 1031, 5.3.1, 5.3.1(A)(B), 5.3.5, and 5.3.5(A)(B)
IFSTA, *Fire Inspection and Code Enforcement,* 6th Edition, page 148.
Delmar, *Fire Prevention, Inspection, and Code Enforcement,* 2nd Edition, page 110.
Answer: C

16. Reference: NFPA 1031, 5.3.10 and 5.3.10(A)(B)
IFSTA, *Fire Inspection and Code Enforcement,* 6th Edition, pages 71–72.
Answer: D

17. Reference: NFPA 1031, 5.3.10 and 5.3.10(A)(B)
IFSTA, *Fire Inspection and Code Enforcement,* 6th Edition, page 59.
Delmar, *Fire Prevention, Inspection, and Code Enforcement,* 2nd Edition, page 123.
Answer: C

18. Reference: NFPA 1031, 5.3.10 and 5.3.10(A)(B)
IFSTA, *Fire Inspection and Code Enforcement,* 6th Edition, page 59.
Delmar, *Fire Prevention, Inspection, and Code Enforcement,* 2nd Edition, page 123.
Answer: D

19. Reference: NFPA 1031, 5.3.11 and 5.3.11(A)(B)
IFSTA, *Fire Inspection and Code Enforcement,* 6th Edition, pages 136–137.
Delmar, *Fire Prevention, Inspection, and Code Enforcement,* 2nd Edition, page 147.
Answer: C

20. Reference: NFPA 1031, 5.3.11 and 5.3.11(A)(B)
IFSTA, *Fire Inspection and Code Enforcement,* 6th Edition, pages 136–137.
Answer: D

21. Reference: NFPA 1031, 5.3.11 and 5.3.11(A)(B)
IFSTA, *Fire Inspection and Code Enforcement,* 6th Edition, page 77.
Answer: B

22. Reference: NFPA 1031, 5.3.12 and 5.3.12(A)(B)
IFSTA, *Fire Inspection and Code Enforcement,* 6th Edition, page 301.
Answer: D

23. Reference: NFPA 1031, 5.3.12, 5.3.12(A)(B), 5.4.4, 5.4.4(A)(B), 5.4.1 and 5.4.1(A)(B)
IFSTA, *Fire Inspection and Code Enforcement,* 6th Edition, page 301.
Answer: C

24. Reference: NFPA 1031, 5.3.2, 5.3.2(A)(B), 5.4.5, and 5.4.5(A)(B)
IFSTA, *Fire Inspection and Code Enforcement,* 6th Edition, page 115.
Delmar, *Fire Prevention, Inspection, and Code Enforcement,* pages 50–52.
Answer: C

25. Reference: NFPA 1031, 5.3.3, 5.3.3(A)(B), 5.3.4, and 5.3.4(A)(B)
IFSTA, *Fire Inspection and Code Enforcement,* 6th Edition, page 141.
Delmar, *Fire Prevention, Inspection, and Code Enforcement,* 2nd Edition, page 111.
Answer: C

26. Reference: NFPA 1031, 5.3.3, 5.3.3(A)(B), 5.4.4, 5.4.4(A)(B), 5.3.5, and 5.3.5(A)(B)
IFSTA, *Fire Inspection and Code Enforcement,* 6th Edition, page 141.
Answer: A

27. Reference: NFPA 1031, 5.3.4 and 5.3.4(A)(B)
IFSTA, *Fire Inspection and Code Enforcement,* 6th Edition, page 160.
Answer: C

28. Reference: NFPA 1031, 5.3.4, 5.3.4(A)(B), 5.4.3, and 5.4.3(A)(B)
IFSTA, *Fire Inspection and Code Enforcement,* 6th Edition, page 248.
Answer: D

29. Reference: NFPA 1031, 5.3.4, 5.3.4(A)(B), 5.4.3, and 5.4.3(A)(B)
IFSTA, *Fire Inspection and Code Enforcement,* 6th Edition, page 173.
Delmar, *Fire Prevention, Inspection, and Code Enforcement,* 2nd Edition, page 83.
Answer: C

30. Reference: NFPA 1031, 5.3.4 and 5.3.4(A)(B)
IFSTA, *Fire Inspection and Code Enforcement,* 6th Edition, page 167.
Answer: C

31. Reference: NFPA 1031, 5.3.4, 5.3.4(A)(B), 5.4.3, and 5.4.3(A)(B)
IFSTA, *Fire Inspection and Code Enforcement,* 6th Edition, page 248.
Answer: A

32. Reference: NFPA 1031, 5.3.4, 5.3.4(A)(B), 5.4.3, 5.4.3(A)(B), 5.3.12, and 5.3.12(A)(B)
IFSTA, *Fire Inspection and Code Enforcement,* 6th Edition, page 207.
Answer: B

33. Reference: NFPA 1031, 5.3.4, 5.3.4(A)(B), 5.4.3, and 5.4.3(A)(B)
IFSTA, *Fire Inspection and Code Enforcement,* 6th Edition, pages 206–207.
Answer: A

34. Reference: NFPA 1031, 5.3.4, 5.3.4(A)(B), 5.4.3, and 5.4.3(A)(B)
IFSTA, *Fire Inspection and Code Enforcement,* 6th Edition, page 207.
Answer: A

35. Reference: NFPA 1031, 5.3.4, 5.3.4(A)(B), 5.4.3, and 5.4.3(A)(B)
IFSTA, *Fire Inspection and Code Enforcement,* 6th Edition, page 207.
Delmar, *Fire Prevention, Inspection, and Code Enforcement,* 2nd Edition, page 88.
Answer: C

36. Reference: NFPA 1031, 5.3.4, 5.3.4(A)(B), 5.4.3, and 5.4.3(A)(B)
IFSTA, *Fire Inspection and Code Enforcement,* 6th Edition, page 190.
Answer: C

37. Reference: NFPA 1031, 5.3.4, 5.3.4(A)(B), 5.4.3, and 5.4.3(A)(B)
IFSTA, *Fire Inspection and Code Enforcement,* 6th Edition, page 190.
Answer: B

38. Reference: NFPA 1031, 5.3.4, 5.3.4(A)(B), 5.4.3, and 5.4.3(A)(B)
IFSTA, *Fire Inspection and Code Enforcement,* 6th Edition, pages 238–242.
Answer: D

39. Reference: NFPA 1031, 5.3.4, 5.3.4(A)(B), 5.4.3, and 5.4.3(A)(B)
IFSTA, *Fire Inspection and Code Enforcement,* 6th Edition, page 162, Table 7.1.
Answer: B

40. Reference: NFPA 1031, 5.3.4, 5.3.4(A)(B), 5.4.3, and 5.4.3(A)(B)
IFSTA, *Fire Inspection and Code Enforcement,* 6th Edition, page 175.
Answer: A

41. Reference: NFPA 1031, 5.3.4, 5.3.4(A)(B), 5.4.3, and 5.4.3(A)(B)
IFSTA, *Fire Inspection and Code Enforcement,* 6th Edition, page 167.
Answer: D

42. Reference: NFPA 1031, 5.3.5, 5.3.5(A)(B), 5.3.4, 5.3.4(A)(B), 5.4.4, and 5.4.4(A)(B)
IFSTA, *Fire Inspection and Code Enforcement,* 6th Edition, page 141.
Answer: D

43. Reference: NFPA 1031, 5.3.5, 5.3.5(A)(B), 5.4.4, and 5.4.4(A)(B)
IFSTA, *Fire Inspection and Code Enforcement,* 6th Edition, page 148.
Delmar, *Fire Prevention, Inspection, and Code Enforcement,* 2nd Edition, pages 102–110.
Answer: B

44. Reference: NFPA 1031, 5.3.5, 5.3.5(A)(B), 5.4.4, and 5.4.4(A)(B)
IFSTA, *Fire Inspection and Code Enforcement,* 6th Edition, page 138.
Answer: D

45. Reference: NFPA 1031, 5.3.5, 5.3.5(A)(B), 5.3.1, and 5.3.1(A)(B)
IFSTA, *Fire Inspection and Code Enforcement,* 6th Edition, pages 140-141.
Delmar, *Fire Prevention, Inspection, and Code Enforcement,* 2nd Edition, page 110.
Answer: E

46. Reference: NFPA 1031, 5.3.5, 5.3.5(A)(B), 5.4.1, 5.4.1(A)(B), 5.4.4, 5.4.4(A)(B), 5.3.12, and 5.3.12(A)(B)
IFSTA, *Fire Inspection and Code Enforcement,* 6th Edition, page 289.
Answer: C

47. Reference: NFPA 1031, 5.3.6 and 5.3.6(A)(B)
IFSTA, *Fire Inspection and Code Enforcement,* 6th Edition, page 338.
Answer: B

48. Reference: NFPA 1031, 5.3.6 and 5.3.6(A)(B)
IFSTA, *Fire Inspection and Code Enforcement,* 6th Edition, page 77.
Answer: B

49. Reference: NFPA 1031, 5.3.6, 5.3.6(A)(B), 5.3.8, and 5.3.8(A)(B)
IFSTA, *Fire Inspection and Code Enforcement,* 6th Edition, page 331.
Answer: C

50. Reference: NFPA 1031, 5.3.7 and 5.3.7(A)(B)
IFSTA, *Fire Inspection and Code Enforcement,* 6th Edition, page 155.
Delmar, *Fire Prevention, Inspection, and Code Enforcement,* 2nd Edition, pages 180–181.
Answer: D

51. Reference: NFPA 1031, 5.3.7 and 5.3.7(A)(B)
IFSTA, *Fire Inspection and Code Enforcement,* 6th Edition, page 154.
Delmar, *Fire Prevention, Inspection, and Code Enforcement,* 2nd Edition, pages 181 and 182.
Answer: B

52. Reference: NFPA 1031, 5.3.8 and 5.3.8(A)(B)
IFSTA, *Fire Inspection and Code Enforcement,* 6th Edition, page 338.
Answer: B

53. Reference: NFPA 1031, 5.3.8, 5.3.8(A)(B), 5.3.6, and 5.3.6(A)(B)
IFSTA, *Fire Inspection and Code Enforcement,* 6th Edition, pages 335–336.
Answer: A

54. Reference: NFPA 1031, 5.3.8 and 5.3.8(A)(B)
IFSTA, *Fire Inspection and Code Enforcement,* 6th Edition, page 334.
Delmar, *Fire Prevention, Inspection, and Code Enforcement,* 2nd Edition, page 203.
Answer: D

55. Reference: NFPA 1031, 5.3.8 and 5.3.8(A)(B)
IFSTA, *Fire Inspection and Code Enforcement,* 6th Edition, page 332.
Answer: A

56. Reference: NFPA 1031, 5.3.8 and 5.3.8(A)(B)
IFSTA, *Fire Inspection and Code Enforcement,* 6th Edition, page 332.
Answer: D

57. Reference: NFPA 1031, 5.3.8, 5.3.8(A)(B), 5.3.6, and 5.3.6(A)(B)
IFSTA, *Fire Inspection and Code Enforcement,* 6th Edition, page 326.
Answer: A

58. Reference: NFPA 1031, 5.3.9, 5.3.9(A)(B), 5.3.8, and 5.3.8(A)(B)
IFSTA, *Fire Inspection and Code Enforcement,* 6th Edition, page 317.
Delmar, *Fire Prevention, Inspection, and Code Enforcement,* 2nd Edition, page 194.
Answer: A

59. Reference: NFPA 1031, 5.3.9 and 5.3.9(A)(B)
IFSTA, *Fire Inspection and Code Enforcement,* 6th Edition, pages 315-317.
Delmar, *Fire Prevention, Inspection, and Code Enforcement,* 2nd Edition, page 186.
Answer: C

60. Reference: NFPA 1031, 5.3.9, 5.3.9(A)(B), 5.3.8 and 5.3.8(A)(B)
IFSTA, *Fire Inspection and Code Enforcement,* 6th Edition, page 317.
Answer: D

61. Reference: NFPA 1031, 5.4.1, 5.4.1(A)(B), 5.3.2, and 5.3.2(A)(B)
IFSTA, *Fire Inspection and Code Enforcement,* 6th Edition, page 124.
Answer: D

62. Reference: NFPA 1031, 5.4.1, 5.4.1(A)(B), 5.4.5, and 5.4.5(A)(B)
IFSTA, *Fire Inspection and Code Enforcement,* 6th Edition, page 115.
Delmar, *Fire Prevention, Inspection, and Code Enforcement,* 2nd Edition, page 50.
Answer: A

63. Reference: NFPA 1031, 5.4.1 and 5.4.1(A)(B)
IFSTA, *Fire Inspection and Code Enforcement,* 6th Edition, page 289.
Answer: C

64. Reference: NFPA 1031, 5.4.1, 5.4.1(A)(B), 5.3.7, and 5.3.7(A)(B)
IFSTA, *Fire Inspection and Code Enforcement,* 6th Edition, page 292.
Answer: B

65. Reference: NFPA 1031, 5.4.2, 5.4.2(A)(B), 5.3.1, and 5.3.1(A)(B)
IFSTA, *Fire Inspection and Code Enforcement,* 6th Edition, page 146.
Delmar, *Fire Prevention, Inspection, and Code Enforcement,* 2nd Edition, pages 103–105.
Answer: D

66. Reference: NFPA 1031, 5.4.3 and 5.4.3(A)(B)
IFSTA, *Fire Inspection and Code Enforcement,* 6th Edition, page 207.
Answer: D

67. Reference: NFPA 1031, 5.4.3, 5.4.3(A)(B), 5.3.4, and 5.3.4(A)(B)
IFSTA, *Fire Inspection and Code Enforcement,* 6th Edition, page 209.
Answer: B

68. Reference: NFPA 1031, 5.4.3, 5.4.3(A)(B), 5.3.4, and 5.3.4(A)(B)
IFSTA, *Fire Inspection and Code Enforcement,* 6th Edition, pages 206–207.
Answer: D

69. Reference: NFPA 1031, 5.4.5, 5.4.5(A)(B), 5.3.3, and 5.3.3(A)(B)
IFSTA, *Fire Inspection and Code Enforcement,* 6th Edition, page 58.
Answer: A

70. Reference: NFPA 1031, 5.4.5, 5.4.5(A)(B), 5.3.3, and 5.3.3(A)(B)
IFSTA, *Fire Inspection and Code Enforcement,* 6th Edition, page 64.
Answer: A

71. Reference: NFPA 1031, 5.4.5 and 5.4.5(A)(B)
IFSTA, *Fire Inspection and Code Enforcement,* 6th Edition, page 57.
Answer: E

72. Reference: NFPA 1031, 5.4.5, 5.4.5(A)(B), 5.3.3, and 5.3.3(A)(B)
IFSTA, *Fire Inspection and Code Enforcement,* 6th Edition, page 116.
Answer: A

73. Reference: NFPA 1031, 5.4.5 and 5.4.5(A)(B)
IFSTA, *Fire Inspection and Code Enforcement,* 6th Edition, pages 62–63.
Answer: C

74. Reference: NFPA 1031, 5.4.5 and 5.4.5(A)(B)
IFSTA, *Fire Inspection and Code Enforcement,* 6th Edition, page 117.
Delmar, *Fire Prevention, Inspection, and Code Enforcement,* 2nd Edition, page 51.
Answer: D

75. Reference: NFPA 1031, 5.4.5 and 5.4.5(A)(B)
IFSTA, *Fire Inspection and Code Enforcement,* 6th Edition, page 63.
Answer: C

Examination II-3 Answer Key

Directions

Follow these steps carefully for completing the feedback part of the Systematic Approach to Examination Preparation (SAEP):

1. After entering your scores, look up the answers for the examination items you missed as well as those you guessed, even if you guessed correctly. If you are guessing, it means the answer isn't perfectly clear. This process will make you as knowledgeable as possible.
2. Enter the number of missed and guessed examination items in the blank on your Personal Progress Plotter.
3. Highlight the answer in the reference materials, and then read the paragraph preceding and the paragraph following the one in which the correct answer is located. Enter the paragraph number and page number next to the guessed or missed examination item on your examination. Count any part of a paragraph at the beginning of the page as one paragraph until you reach the paragraph containing your highlighted answer. This step will help you locate and review your missed and guessed examination items later in the process. This step is essential to learning the material in context and by association. These learning techniques (context/association) are the very backbone of the SAEP approach.
4. **Congratulations**! You have completed the examination and feedback parts of SAEP when you have highlighted your guessed and missed examination items for this examination.

Proceed to Phase III and Phase IV. Study the materials carefully in these important phases. They will help you polish your examination-taking skills. Approximately two to three days prior to taking your next examination, carefully read all of the highlighted information in the reference materials using the same techniques applied during the feedback part. This exercise will reinforce your learning and provide you with an added level of confidence going into the examination.

Someone once said to professional golfer Tom Watson after he won several tournament championships, "You are really lucky to have won those championships. You are really on a streak." Watson was reported to have replied, "Yes, there is some luck involved, but what I've really noticed is that the more I practice, the luckier I get." What Watson was saying is that good luck usually results from good preparation. This line of thinking certainly applies to learning the rules and hints of examination taking.

Rule 7

Good luck = Good preparation.

1. Reference: NFPA 1031, 5.2
   IFSTA, *Fire Inspection and Code Enforcement,* 6th Edition, pages 13–14.
   Answer: B

2. Reference: NFPA 1031, 5.2
   IFSTA, *Fire Inspection and Code Enforcement,* 6th Edition, page 13.
   Answer: D

3. Reference: NFPA 1031, 5.2
IFSTA, *Fire Inspection and Code Enforcement,* 6th Edition, page 14.
Answer: C

4. Reference: NFPA 1031, 5.2
IFSTA, *Fire Inspection and Code Enforcement,* 6th Edition, page 6.
Delmar, *Fire Prevention, Inspection, and Code Enforcement,* 2nd Edition, page 28.
Answer: B

5. Reference: NFPA 1031, 5.2.1, 5.2.1(A)(B), 5.2.4, and 5.2.4(A)(B)
IFSTA, *Fire Inspection and Code Enforcement,* 6th Edition, pages 18–19.
Answer: B

6. Reference: NFPA 1031, 5.2.1 and 5.2.1(A)(B)
IFSTA, *Fire Inspection and Code Enforcement,* 6th Edition, page 20.
Answer: B

7. Reference: NFPA 1031, 5.2.2 and 5.2.2(A)(B)
IFSTA, *Fire Inspection and Code Enforcement,* 6th Edition, page 289.
Answer: A

8. Reference: NFPA 1031, 5.2.2 and 5.2.2(A)(B)
IFSTA, *Fire Inspection and Code Enforcement,* 6th Edition, page 289.
Answer: C

9. Reference: NFPA 1031, 5.2.2 and 5.2.2(A)(B)
IFSTA, *Fire Inspection and Code Enforcement,* 6th Edition, page 289.
Answer: D

10. Reference: NFPA 1031, 5.2.2, 5.2.2(A)(B), 5.4.1, and 5.4.1(A)(B)
IFSTA, *Fire Inspection and Code Enforcement,* 6th Edition, pages 289 and 452.
Answer: C

11. Reference: NFPA 1031, 5.2.3 and 5.2.3(A)(B)
IFSTA, *Fire Inspection and Code Enforcement,* 6th Edition, pages 14–15.
Answer: C

12. Reference: NFPA 1031, 5.2.3 and 5.2.3(A)(B)
IFSTA, *Fire Inspection and Code Enforcement,* 6th Edition, pages 14–15.
Answer: D

13. Reference: NFPA 1031, 5.2.3 and 5.2.3(A)(B)
IFSTA, *Fire Inspection and Code Enforcement,* 6th Edition, pages 14–15.
Answer: B

14. Reference: NFPA 1031, 5.2.4 and 5.2.4(A)(B)
IFSTA, *Fire Inspection and Code Enforcement,* 6th Edition, pages 11–12.
Answer: C

15. Reference: NFPA 1031, 5.2.4 and 5.2.4(A)(B)
IFSTA, *Fire Inspection and Code Enforcement,* 6th Edition, pages 11–12.
Answer: D

16. Reference: NFPA 1031, 5.2.5 and 5.2.5(A)(B)
IFSTA, *Fire Inspection and Code Enforcement,* 6th Edition, page 36.
Answer: A

17. Reference: NFPA 1031, 5.2.5 and 5.2.5(A)(B)
IFSTA, *Fire Inspection and Code Enforcement,* 6th Edition, page 36.
Answer: B

18. Reference: NFPA 1031, 5.2.5 and 5.2.5(A)(B)
IFSTA, *Fire Inspection and Code Enforcement,* 6th Edition, page 36.
Answer: C

19. Reference: NFPA 1031, 5.2.5 and 5.2.5(A)(B)
IFSTA, *Fire Inspection and Code Enforcement,* 6th Edition, page 36.
Answer: C

20. Reference: NFPA 1031, 5.2.5, 5.2.5(A)(B), 5.2.4, and 5.2.4(A)(B)
IFSTA, *Fire Inspection and Code Enforcement,* 6th Edition, pages 379–380.
Answer: D

21. Reference: NFPA 1031, 5.2.5 and 5.2.5(A)(B)
IFSTA, *Fire Inspection and Code Enforcement,* 6th Edition, pages 33–34.
Answer: C

22. Reference: NFPA 1031, 5.2.5 and 5.2.5(A)(B)
IFSTA, *Fire Inspection and Code Enforcement,* 6th Edition, page 33.
Answer: B

23. Reference: NFPA 1031, 5.2.5, 5.2.5(A)(B), 5.2.3, and 5.2.3(A)(B)
IFSTA, *Fire Inspection and Code Enforcement,* 6th Edition, pages 5–6.
Answer: D

24. Reference: NFPA 1031, 5.2.5 and 5.2.5(A)(B)
IFSTA, *Fire Inspection and Code Enforcement,* 6th Edition, pages 5–6.
Answer: D

25. Reference: NFPA 1031, 5.3.1, 5.3.1(A)(B), 5.3.3, and 5.3.3(A)(B)
IFSTA, *Fire Inspection and Code Enforcement,* 6th Edition, pages 146–150.
Answer: A

26. Reference: NFPA 1031, 5.3.1, 5.3.1(A)(B), 5.3.5, and 5.3.5(A)(B)
IFSTA, *Fire Inspection and Code Enforcement,* 6th Edition, page 148.
Delmar, *Fire Prevention, Inspection, and Code Enforcement,* 2nd Edition, page 110.
Answer: C

27. Reference: NFPA 1031, 5.3.1, 5.3.1(A)(B), 5.3.5, 5.3.5(A)(B), 5.4.4, and 5.4.4(A)(B)
IFSTA, *Fire Inspection and Code Enforcement,* 6th Edition, page 150.
Delmar, *Fire Prevention, Inspection, and Code Enforcement,* 2nd Edition, page 106.
Answer: B

28. Reference: NFPA 1031, 5.3.10 and 5.3.10(A)(B)
IFSTA, *Fire Inspection and Code Enforcement,* 6th Edition, page 71.
Answer: B

29. Reference: NFPA 1031, 5.3.10 and 5.3.10(A)(B)
IFSTA, *Fire Inspection and Code Enforcement,* 6th Edition, pages 59–60.
Delmar, *Fire Prevention, Inspection, and Code Enforcement,* 2nd Edition, page 123.
Answer: D

30. Reference: NFPA 1031, 5.3.10 and 5.3.10(A)(B)
IFSTA, *Fire Inspection and Code Enforcement,* 6th Edition, pages 71–72.
Answer: D

31. Reference: NFPA 1031, 5.3.10 and 5.3.10(A)(B)
IFSTA, *Fire Inspection and Code Enforcement,* 6th Edition, page 59.
Delmar, *Fire Prevention, Inspection, and Code Enforcement,* 2nd Edition, page 123.
Answer: C

32. Reference: NFPA 1031, 5.3.10 and 5.3.10(A)(B)
IFSTA, *Fire Inspection and Code Enforcement,* 6th Edition, page 59.
Delmar, *Fire Prevention, Inspection, and Code Enforcement,* 2nd Edition, page 123.
Answer: D

33. Reference: NFPA 1031, 5.3.10 and 5.3.10(A)(B)
IFSTA, *Fire Inspection and Code Enforcement,* 6th Edition, page 57.
Delmar, *Fire Prevention, Inspection, and Code Enforcement,* 2nd Edition, page 63.
Answer: B

34. Reference: NFPA 1031, 5.3.11 and 5.3.11(A)(B)
IFSTA, *Fire Inspection and Code Enforcement,* 6th Edition, page 136.
Delmar, *Fire Prevention, Inspection, and Code Enforcement,* 2nd Edition, page 147.
Answer: C

35. Reference: NFPA 1031, 5.3.11 and 5.3.11(A)(B)
IFSTA, *Fire Inspection and Code Enforcement,* 6th Edition, page 136.
Answer: A

36. Reference: NFPA 1031, 5.3.11 and 5.3.11(A)(B)
IFSTA, *Fire Inspection and Code Enforcement,* 6th Edition, page 135.
Answer: B

37. Reference: NFPA 1031, 5.3.11 and 5.3.11(A)(B)
IFSTA, *Fire Inspection and Code Enforcement,* 6th Edition, page 136.
Answer: C

38. Reference: NFPA 1031, 5.3.12 and 5.3.12(A)(B)
IFSTA, *Fire Inspection and Code Enforcement,* 6th Edition, page 301.
Answer: D

39. Reference: NFPA 1031, 5.3.12, 5.3.12(A)(B), 5.4.4, 5.4.4(A)(B), 5.4.1, and 5.4.1(A)(B)
IFSTA, *Fire Inspection and Code Enforcement,* 6th Edition, page 301.
Answer: C

40. Reference: NFPA 1031, 5.3.2, 5.3.2(A)(B), 5.4.1, 5.4.1(A)(B), 5.4.2, 5.4.2(A)(B), 5.4.5, 5.4.5(A)(B), 5.3.12, and 5.3.12(A)(B)
IFSTA, *Fire Inspection and Code Enforcement,* 6th Edition, page 115.
Answer: D

41. Reference: NFPA 1031, 5.3.3, 5.3.3(A)(B), 5.4.4, 5.4.4(A)(B), 5.3.5, and 5.3.5(A)(B)
IFSTA, *Fire Inspection and Code Enforcement,* 6th Edition, page 141.
Answer: A

42. Reference: NFPA 1031, 5.3.4 and 5.3.4(A)(B)
IFSTA, *Fire Inspection and Code Enforcement,* 6th Edition, page 249.
Answer: C

43. Reference: NFPA 1031, 5.3.4 and 5.3.4(A)(B)
IFSTA, *Fire Inspection and Code Enforcement,* 6th Edition, pages 249–257.
Answer: A

44. Reference: NFPA 1031, 5.3.4, 5.3.4(A)(B), 5.4.3, and 5.4.3(A)(B)
IFSTA, *Fire Inspection and Code Enforcement,* 6th Edition, page 452.
Answer: B

45. Reference: NFPA 1031, 5.3.4, 5.3.4(A)(B), 5.4.3, and 5.4.3(A)(B)
IFSTA, *Fire Inspection and Code Enforcement,* 6th Edition, page 164.
Answer: D

46. Reference: NFPA 1031, 5.3.4 and 5.3.4(A)(B)
IFSTA, *Fire Inspection and Code Enforcement,* 6th Edition, page 210.
Answer: C

47. Reference: NFPA 1031, 5.3.4 and 5.3.4(A)(B)
IFSTA, *Fire Inspection and Code Enforcement,* 6th Edition, page 210.
Answer: B

48. Reference: NFPA 1031, 5.3.4 and 5.3.4(A)(B)
IFSTA, *Fire Inspection and Code Enforcement,* 6th Edition, page 160.
Answer: C

49. Reference: NFPA 1031, 5.3.4, 5.3.4(A)(B), 5.4.3, and 5.4.3(A)(B)
IFSTA, *Fire Inspection and Code Enforcement,* 6th Edition, page 248.
Answer: D

50. Reference: NFPA 1031, 5.3.4, 5.3.4(A)(B), 5.4.3, and 5.4.3(A)(B)
IFSTA, *Fire Inspection and Code Enforcement,* 6th Edition, page 248.
Answer: A

51. Reference: NFPA 1031, 5.3.4, 5.3.4(A)(B), 5.4.3, 5.4.3(A)(B), 5.3.12, and 5.3.12(A)(B)
IFSTA, *Fire Inspection and Code Enforcement,* 6th Edition, page 207.
Answer: B

52. Reference: NFPA 1031, 5.3.4, 5.3.4(A)(B), 5.4.3, and 5.4.3(A)(B)
IFSTA, *Fire Inspection and Code Enforcement,* 6th Edition, pages 206–207.
Answer: A

53. Reference: NFPA 1031, 5.3.4, 5.3.4(A)(B), 5.4.3, and 5.4.3(A)(B)
IFSTA, *Fire Inspection and Code Enforcement,* 6th Edition, page 207.
Answer: A

54. Reference: NFPA 1031, 5.3.4, 5.3.4(A)(B), 5.4.3, and 5.4.3(A)(B)
IFSTA, *Fire Inspection and Code Enforcement,* 6th Edition, page 190.
Answer: B

55. Reference: NFPA 1031, 5.3.4, 5.3.4(A)(B), 5.4.3, and 5.4.3(A)(B)
IFSTA, *Fire Inspection and Code Enforcement,* 6th Edition, pages 238–242.
Answer: D

56. Reference: NFPA 1031, 5.3.4, 5.3.4(A)(B), 5.4.3, 5.4.3(A)(B), 5.3.12, 5.3.12(A)(B), 5.4.4, and 5.4.4(A)(B)
IFSTA, *Fire Inspection and Code Enforcement,* 6th Edition, page 301.
Answer: D

57. Reference: NFPA 1031, 5.3.4, 5.3.4(A)(B), 5.4.3, and 5.4.3(A)(B)
IFSTA, *Fire Inspection and Code Enforcement,* 6th Edition, pages 176–178.
Answer: B

58. Reference: NFPA 1031, 5.3.4, 5.3.4(A)(B), 5.4.3, and 5.4.3(A)(B)
IFSTA, *Fire Inspection and Code Enforcement,* 6th Edition, page 174.
Answer: B

59. Reference: NFPA 1031, 5.3.4, 5.3.4(A)(B), 5.4.3, and 5.4.3(A)(B)
IFSTA, *Fire Inspection and Code Enforcement,* 6th Edition, page 171.
Answer: C

60. Reference: NFPA 1031, 5.3.4, 5.3.4(A)(B), 5.4.3, and 5.4.3(A)(B)
IFSTA, *Fire Inspection and Code Enforcement,* 6th Edition, page 449.
Answer: B

61. Reference: NFPA 1031, 5.3.4, 5.3.4(A)(B), 5.4.3, and 5.4.3(A)(B)
IFSTA, *Fire Inspection and Code Enforcement,* 6th Edition, page 163.
Answer: B

62. Reference: NFPA 1031, 5.3.4, 5.3.4(A)(B), 5.4.3, 5.4.3(A)(B), 5.3.12, and 5.3.12(A)(B)
IFSTA, *Fire Inspection and Code Enforcement,* 6th Edition, page 167.
Answer: B

63. Reference: NFPA 1031, 5.3.4, 5.3.4(A)(B), 5.4.3, and 5.4.3(A)(B)
IFSTA, *Fire Inspection and Code Enforcement,* 6th Edition, pages 176–177.
Answer: B

64. Reference: NFPA 1031, 5.3.4, 5.3.4(A)(B), 5.4.3, and 5.4.3(A)(B)
IFSTA, *Fire Inspection and Code Enforcement,* 6th Edition, page 171.
Answer: B

65. Reference: NFPA 1031, 5.3.4, 5.3.4(A)(B), 5.4.3, 5.4.3(A)(B), 5.3.12, and 5.3.12(A)(B)
IFSTA, *Fire Inspection and Code Enforcement,* 6th Edition, page 162, Table 7.1.
Answer: C

66. Reference: NFPA 1031, 5.3.4, 5.3.4(A)(B), 5.4.3, 5.4.3(A)(B), 5.3.12, and 5.3.12(A)(B)
IFSTA, *Fire Inspection and Code Enforcement,* 6th Edition, page 454.
Answer: D

67. Reference: NFPA 1031, 5.3.5 and 5.3.5(A)(B)
IFSTA, *Fire Inspection and Code Enforcement,* 6th Edition, page 139.
Answer: A

68. Reference: NFPA 1031, 5.3.5, 5.3.5(A)(B), 5.3.4, and 5.3.4(A)(B)
IFSTA, *Fire Inspection and Code Enforcement,* 6th Edition, page 144.
Answer: A

69. Reference: NFPA 1031, 5.3.5, 5.3.5(A)(B), 5.4.1, 5.4.1(A)(B), 5.4.4, 5.4.4(A)(B), 5.3.12, and 5.3.12(A)(B)
IFSTA, *Fire Inspection and Code Enforcement,* 6th Edition, page 297.
Answer: B

70. Reference: NFPA 1031, 5.3.5, 5.3.5(A)(B), 5.4.4, and 5.4.4(A)(B)
IFSTA, *Fire Inspection and Code Enforcement,* 6th Edition, page 148.
Delmar, *Fire Prevention, Inspection, and Code Enforcement,* 2nd Edition, pages 102–110.
Answer: B

71. Reference: NFPA 1031, 5.3.5, 5.3.5(A)(B), 5.4.1, 5.4.1(A)(B), 5.4.4, 5.4.4(A)(B), 5.3.12, and 5.3.12(A)(B)
IFSTA, *Fire Inspection and Code Enforcement,* 6th Edition, page 289.
Answer: C

72. Reference: NFPA 1031, 5.3.6 and 5.3.6(A)(B)
IFSTA, *Fire Inspection and Code Enforcement,* 6th Edition, page 93.
Answer: D

73. Reference: NFPA 1031, 5.3.6 and 5.3.6(A)(B)
IFSTA, *Fire Inspection and Code Enforcement,* 6th Edition, pages 88–90.
Answer: B

74. Reference: NFPA 1031, 5.3.6, 5.3.6(A)(B), 5.3.8, and 5.3.8(A)(B)
IFSTA, *Fire Inspection and Code Enforcement,* 6th Edition, page 338.
Answer: A

75. Reference: NFPA 1031, 5.3.6, 5.3.6(A)(B), 5.4.1, and 5.4.1(A)(B)
IFSTA, *Fire Inspection and Code Enforcement,* 6th Edition, page 300.
Answer: C

76. Reference: NFPA 1031, 5.3.7 and 5.3.7(A)(B)
IFSTA, *Fire Inspection and Code Enforcement,* 6th Edition, page 153.
Delmar, *Fire Prevention, Inspection, and Code Enforcement,* 2nd Edition, page 181.
Answer: A

77. Reference: NFPA 1031, 5.3.7 and 5.3.7(A)(B)
IFSTA, *Fire Inspection and Code Enforcement,* 6th Edition, page 154.
Delmar, *Fire Prevention, Inspection, and Code Enforcement,* 2nd Edition, pages 181–182.
Answer: B

78. Reference: NFPA 1031, 5.3.8, 5.3.8(A)(B), 5.3.9, and 5.3.9(A)(B)
IFSTA, *Fire Inspection and Code Enforcement,* 6th Edition, page 350.
Answer: C

79. Reference: NFPA 1031, 5.3.8 and 5.3.8(A)(B)
IFSTA, *Fire Inspection and Code Enforcement,* 6th Edition, pages 325–326.
Delmar, *Fire Prevention, Inspection, and Code Enforcement,* 2nd Edition, pages 202–203.
Answer: D

80. Reference: NFPA 1031, 5.3.8, 5.3.8(A)(B), 5.3.6, and 5.3.6(A)(B)
IFSTA, *Fire Inspection and Code Enforcement,* 6th Edition, pages 335–336.
Answer: A

81. Reference: NFPA 1031, 5.3.8, 5.3.8(A)(B), 5.3.6, and 5.3.6(A)(B)
IFSTA, *Fire Inspection and Code Enforcement,* 6th Edition, page 326.
Answer: A

82. Reference: NFPA 1031, 5.3.8, 5.3.8(A)(B), 5.3.6, and 5.3.6(A)(B)
IFSTA, *Fire Inspection and Code Enforcement,* 6th Edition, page 350.
Answer: A

83. Reference: NFPA 1031, 5.3.8, 5.3.8(A)(B), 5.3.9, and 5.3.9(A)(B)
IFSTA, *Fire Inspection and Code Enforcement,* 6th Edition, page 339.
Answer: D

84. Reference: NFPA 1031, 5.3.9, 5.3.9(A)(B), 5.3.8, and 5.3.8(A)(B)
IFSTA, *Fire Inspection and Code Enforcement,* 6th Edition, page 317.
Delmar, *Fire Prevention, Inspection, and Code Enforcement,* 2nd Edition, page 194.
Answer: A

85. Reference: NFPA 1031, 5.4.1, 5.4.1(A)(B), 5.3.2 and 5.3.2(A)(B)
IFSTA, *Fire Inspection and Code Enforcement,* 6th Edition, page 131.
Answer: A

86. Reference: NFPA 1031, 5.4.1, and 5.4.1(A)(B)
IFSTA, *Fire Inspection and Code Enforcement,* 6th Edition, page 122.
Answer: B

87. Reference: NFPA 1031, 5.4.1, 5.4.1(A)(B), 5.4.5, and 5.4.5(A)(B)
IFSTA, *Fire Inspection and Code Enforcement,* 6th Edition, page 115.
Delmar, *Fire Prevention, Inspection, and Code Enforcement,* 2nd Edition, page 50.
Answer: A

88. Reference: NFPA 1031, 5.4.1, 5.4.1(A)(B), 5.3.7, and 5.3.7(A)(B)
IFSTA, *Fire Inspection and Code Enforcement,* 6th Edition, pages 290–292.
Answer: A

89. Reference: NFPA 1031, 5.4.2, and 5.4.2(A)(B), 5.3.5, 5.3.5(A)(B), 5.3.1, and 5.3.1(A)(B)
IFSTA, *Fire Inspection and Code Enforcement,* 6th Edition, page 150.
Delmar, *Fire Prevention, Inspection, and Code Enforcement,* 2nd Edition, page 110.
Answer: C

90. Reference: NFPA 1031, 5.4.2, 5.4.2(A)(B), 5.3.1, and 5.3.1(A)(B)
IFSTA, *Fire Inspection and Code Enforcement,* 6th Edition, page 146.
Delmar, *Fire Prevention, Inspection, and Code Enforcement,* 2nd Edition, pages 103–105.
Answer: D

91. Reference: NFPA 1031, 5.4.3, 5.4.3(A)(B), 5.3.4, and 5.3.4(A)(B)
IFSTA, *Fire Inspection and Code Enforcement,* 6th Edition, page 226.
Answer: B

92. Reference: NFPA 1031, 5.4.3, 5.4.3(A)(B), 5.3.4, and 5.3.4(A)(B)
IFSTA, *Fire Inspection and Code Enforcement,* 6th Edition, page 209.
Answer: B

93. Reference: NFPA 1031, 5.4.3, 5.4.3(A)(B), 5.3.4, and 5.3.4(A)(B)
IFSTA, *Fire Inspection and Code Enforcement,* 6th Edition, pages 206–207.
Answer: D

94. Reference: NFPA 1031, 5.4.3, 5.4.3(A)(B), 5.3.4, 5.3.4(A)(B), 5.3.12, and 5.3.12(A)(B)
IFSTA, *Fire Inspection and Code Enforcement,* 6th Edition, page 217.
Delmar, *Fire Prevention, Inspection, and Code Enforcement,* 2nd Edition, page 164.
Answer: D

95. Reference: NFPA 1031, 5.4.5 and 5.4.5(A)(B)
IFSTA, *Fire Inspection and Code Enforcement,* 6th Edition, page 67.
Delmar, *Fire Prevention, Inspection, and Code Enforcement,* 2nd Edition, page 69.
Answer: D

96. Reference: NFPA 1031, 5.4.5 and 5.4.5(A)(B)
IFSTA, *Fire Inspection and Code Enforcement,* 6th Edition, pages 58.
Answer: B

97. Reference: NFPA 1031, 5.4.5, 5.4.5(A)(B), 5.3.3, and 5.3.3(A)(B)
IFSTA, *Fire Inspection and Code Enforcement,* 6th Edition, pages 115–116.
Answer: A

98. Reference: NFPA 1031, 5.4.5, 5.4.5(A)(B), 5.3.3, and 5.3.3(A)(B)
IFSTA, *Fire Inspection and Code Enforcement,* 6th Edition, page 64.
Answer: A

99. Reference: NFPA 1031, 5.4.5 and 5.4.5(A)(B)
IFSTA, *Fire Inspection and Code Enforcement,* 6th Edition, pages 62–63.
Answer: C

100. Reference: NFPA 1031, 5.4.5 and 5.4.5(A)(B)
IFSTA, *Fire Inspection and Code Enforcement,* 6th Edition, page 117.
Delmar, *Fire Prevention, Inspection, and Code Enforcement,* 2nd Edition, page 51.
Answer: D

Bibliography for Exam Prep: Fire Inspector I and II

1. NFPA 1031, *Standard for Professional Qualifications for Fire Inspector and Plan Examiner*, 2003 edition.
2. IFSTA, *Fire Inspection and Code Enforcement,* Sixth Edition.
3. Delmar, *Fire Prevention, Inspection, and Code Enforcement,* Second Edition.
4. IFSTA, *Essentials of Fire Fighting,* Fourth Edition.